The Brilliant
Zewail

The Brilliant
Zewail

Editor
Lotfia El-Nadi
Cairo University, Egypt

Book Designer
Mostafa Nabil

World Scientific

NEW JERSEY · LONDON · SINGAPORE · BEIJING · SHANGHAI · HONG KONG · TAIPEI · CHENNAI · TOKYO

Published by

World Scientific Publishing Co. Pte. Ltd.

5 Toh Tuck Link, Singapore 596224

USA office: 27 Warren Street, Suite 401-402, Hackensack, NJ 07601

UK office: 57 Shelton Street, Covent Garden, London WC2H 9HE

British Library Cataloguing-in-Publication Data
A catalogue record for this book is available from the British Library.

THE BRILLIANT ZEWAIL

ISBN 978-981-3275-82-9
ISBN 978-981-3276-70-3 (pbk)

For any available supplementary material, please visit
https://www.worldscientific.com/worldscibooks/10.1142/11145#t=suppl

Desk Editor: Rhaimie Wahap

Typeset by Stallion Press
Email: enquiries@stallionpress.com

THE BRILLIANT
ZEWAIL

PREFACE

The eleven chapters in this book have each been divided into parts. The first chapter was written by Zewail's sister and is meant to enlighten a period of his life that few people may know. It describes the life of Ahmed Zewail; focusing on his childhood, early family life, glimpses of brilliance and eagerness to achieve scientific success.

The following chapters take a more objective view of his life after his breakthrough invention, in which Zewail became the first person in the world to capture, in a few femtoseconds, the exact moment when molecules divide or unite. That moment marked the dawn of a new era for chemistry. The chapters go on to focusing on the ramifications of his work and his persona on a global context and his efforts on strengthening the basics of science and technology in the land of his birth — Egypt. These chapters also express the ideas created by Zewail — together with his collaborators — to test and measure unprecedented thought experiments that follow the molecular behaviors in space and time, which were otherwise undetermined, inventing the four-dimensional electron microscope.

Femtochemistry and femtobiology provide not only the most detailed information about chemical reaction dynamics, but they hold the promise that, one day, researchers could create completely novel materials. For the past years, these significant fields had yielded thousands of published papers in physical chemistry that are highly cited.

Besides his scientific successes, the chapters also give a hint on Zewail's personal attitude and brilliance, describing him as glamorous and having a wonderful character. Readers are urged to go through Chapter 11 in order to see for themselves Zewail's long list of scientific publications and awards.

Zewail's brilliance did not end with his passing, but instead it laid down the prospects for the fine tuning of the motion and reactivity of molecules. Colleagues and students at Caltech are still publishing papers with Zewail's name as co-author in international journals up to 2018. If proved successful, laser-customised chemistry may be developed in the coming decades.

Lotfia El-Nadi
University of Cairo
20 March 2019

ABOUT THE EDITOR

Lotfia El-Nadi is the Vice Director of the International Committee of Scientific Research, National Institute of Laser Enhanced Sciences (NILES), Cairo University. She is also the Vice Director of the International Center of Scientific and Applied Studies for High Density Short Pulse Lasers, NILES, Cairo University since 2006.

El-Nadi obtained her B.Sc. in physics and chemistry from Cairo University; M.Sc. in radiation physics from Birmingham University, and Ph.D. in nuclear physics from Cairo University. Her expansive academic career includes posts at Cairo University, Qatar University, and King Abdul Aziz University in Saudi Arabia.

From 1990 to 1993, she was director of the National Center of Lasers and Applications at Cairo University. Between 1991 and 1994, she was Head of the Physics Department at Cairo University. She has also been a board member of the National Institute of Laser Enhanced Sciences at Cairo University 2005/2008 and 2016/2018.

Summary of Lotfia El-Nadi's achievements

- Elected a member of the International Committee of Ultra Intense Lasers (ICUIL — www.ICUIL.org) at GSI meeting in Germany (2018).
- Awarded the Trophy of the Indian Government, March (2014).
- Awarded the Egyptian Government Medal Wissam of Science and Arts Grade 1 (2013).
- Appointed member of the Advisory Board of Zewail City of Science and Technology (2012).
- Awarded the Academy of Sciences Highest Credit Award of Advanced Sciences (2009).
- Awarded the Ordre des Palmes Académiques from the Prime Minister of France, October (2004).
- Awarded the Cairo University Shield (1994).
- Awarded the Darmstadt University Golden Medal (1994).
- Established the National Institute of Enhanced Laser Sciences at Cairo University (1994).
- Established the Topical Society of Laser Sciences (TSLS — www.tsls.org.eg) (1987).
- Published more than 125 Papers in international and national journals.
- Supervised 72 Phd and MSc theses, two are still in progress.
- Published two books on Lasers and Applications in Arabic (1990 and 1993).
- Edited 2 AIP International Conference Proceedings (2004 and 2006).
- Editor-in-chief of the MTPR conferences online journal www.mtpr.pub, Dec. 2018.
- Organized seven International Conferences on Modern Trends in Physics Research (MTPR) and four International Workshops on Ultrafast Laser Technology and Applications (UFLTA).

CONTENTS

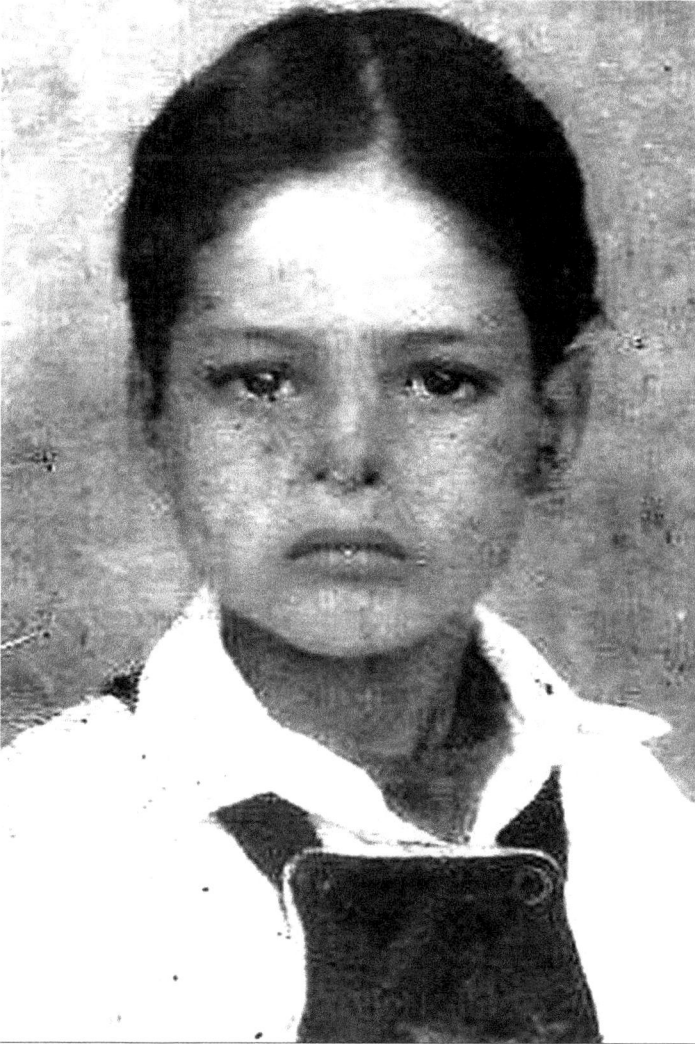

Fig. 1.1

CHAPTER 1

ZEWAIL — THE HISTORIC FINGERPRINT

Nana Zewail

I t might be of interest for readers to know about Zewail's younger years, so in this opening chapter, I will do my best to share with you the most impressionable moments of his life.

1.1 The Childhood of Ahmed Zewail

On a winter day of 1946, during my mother's trip from Dessouque to Damanhur to visit my grandmother, she went into labour and my brother was born. He was named Ahmed after my grandfather, but my mother kept calling him *Shawky*, meaning my desired one, since he was a gift from God after years of longing for a son.

Ahmed Hassan Zewail spent the first 40 days of his life in Damanhur. The years of Ahmed's childhood were spent with his family along the banks of the Nile River Delta of Rosetta Branch in the city of Dessouque — a place blessed with the abundance of the Sun and warm weather — and is reflected in the hearts of the people living there. It was from these people that he derived positive energy and love in his early growing years. Ahmed enjoyed a sheltered, loving and happy childhood surrounded both by his small immediate family and his larger family — the people living along the banks of the Nile River Delta of Rosetta Branch in the city of Dessouque.

Ahmed was a great influence wherever he went, and brought himself up as if to prepare himself for the long journey that would take him to faraway places, places that are often difficult with extreme weather conditions.

Fig. 1.2　On the left side is Mr. Hassan, Ahmed Zewail's father. On the right side is Rawhiaa Dar, his mother.

His immediate family consisted of his parents and three sisters, one of them was me. Our father was a devoted family man, a lover of life; fun-loving and well-loved by everybody. Our mother was characterized by her devotion, religiosity and sternness. She devoted her life to our care and was the driving force that helped Ahmed to excel in his studies. She made sure the family atmosphere was filled with love, stability, security and reassurance.

Our house was close to the grand mosque Ibrahim Desouki, which is located in the heart of the town. The mosque is like a magnet — attracting the people — just like the attracting power of the Sun, where the lives of people revolve around it. At the mosque, you can play simple but entertaining activities. Ahmed loved to play football with his friends there. At home, he would play card games with us. Sometimes we would spend some days of the summer in the hospitality of our relatives in Alexandria, my father's native town. Ahmed enjoyed the beach, the family atmosphere and the grilled fish.

Fig. 1.3   The young Ahmed Zewial with his father on Alexandria beach, 1951.

In school, Ahmed enjoyed the classes and made every effort to achieve success and excellence in his studies. This pleased his family.

In 1952, he joined a school trip to Alexandria University. It was the same year of the 23rd July Revolution. While still a young student, he already had a dream of entering the university. Many Egyptians during that period felt that they were starting a new era of equality among all citizens, especially in opportunities for a free university education.

Fig. 1.4   Ahmed Zewail with his art teacher and fellow students, 1952.

Access to university education was difficult for many poor and even middle-class families. The nationalistic enthusiasm in my young student brother (he was 10 years old at that time — Lord bless him), made him send a letter to the then President Gamal Abdel Nasser, President of the Republic of Egypt in 1956, My brother was delighted to receive a reply from President Gamal Abdel Nasser advising him to maintain enthusiasm in acquiring the deep knowledge of science and to also concentrate on moral attitudes. The letter filled him with enthusiasm to reach the frontiers of science no matter how difficult and to keep in mind the sayings of our prophet, peace be upon him.

He — "Our Prophet" — said, "Take science from the cradle to the grave" He also said, "Seek science, even in China." This event was not the only moment of inspiration in the childhood of Ahmed Zewail. He was raised in an environment that appreciates science, and he knew and have read many times the first Quranic verse revealed to our Prophet Muhammad (peace be upon him) which was:

"In the Name of Allah, the Beneficent, the Merciful. Read (Proclaim!) In the Name of your Lord who created. Created man, out of a clot (of congealed blood). Read (Proclaim), and your Lord is the most generous, who taught by the pen, taught man that which he knew not."

Surah 96 Al 'Alaq, The Quran

My father gave his full blessings to Ahmed's passion for education and diligence and took pride in all of Ahmed's scientific achievements. Not only that, may God's mercy be upon him, he encouraged Ahmed in terms of approbation. The people of the town knew of his excellence in school and he earned respect and love from his friends. Perhaps this is what I spoke to young people about, "Science brings to whoever owns it a high status in the community, and that to achieve one's dream, it requires the owner's creativity, imagination and courage along with perseverance and diligence."

1.2 The Youth Juncture

Ahmed Zewail developed his dream at a very young age. The goal was so clear in his mind that he even wrote it on a piece of paper, "Dr. Ahmed", and stuck it on the door of his room. It was behind this door that he spent hours and hours engaging in serious and difficult studies.

Being the youngest in the family, I witnessed the preparatory stage of my brother. He was 13 years older than me. He did not spare any effort in achieving his dream. He used the summer vacation to read in preparation for next year's semester. His motivation to study was not only due to a sense of duty, but he was well aware that enthusiasm would help him achieve his dream at a faster rate. The long hours of studying were accompanied by songs from "Umm Kulthum". Like any young man, he loves listening to music and he would make the volume on the radio softer so that he could study with the music of his favourite singer in the background.

His deep concentration in his academic pursuits did not prevent him from participating in the various school activities. In school, he joined the basketball team and participated in the drawing group. He

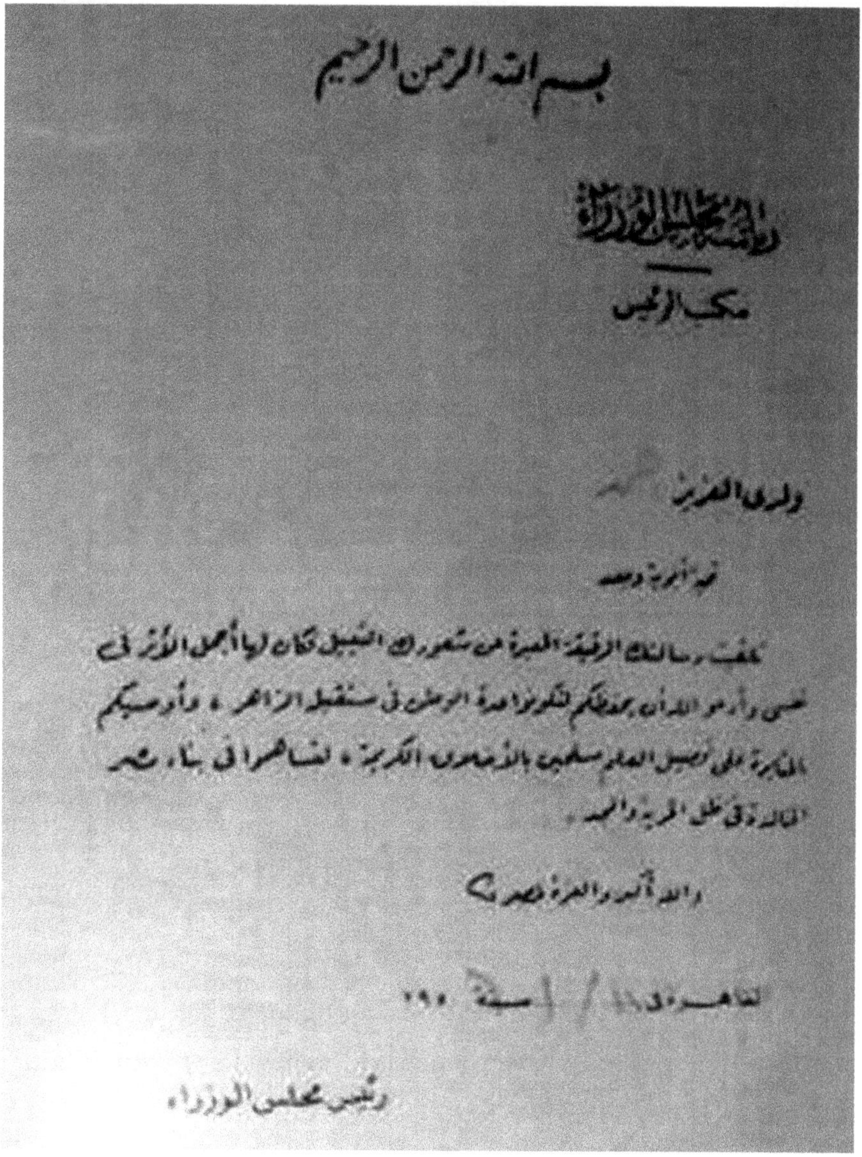

بسم الله الرحمن الرحيم

رئاسة مجلس الوزراء

مكتب الرئيس

ولدى العزيز

تحية طيبة وبعد

تلقيت رسالتك الرقيقة المعبرة عن شعورك النبيل فكان لها أجمل الأثر فى نفسى وأدعو الله أن يوفقكم لتكونوا عدة الوطن فى مستقبله الزاهر ، وأوصيكم بالمثابرة على تحصيل العلم مسلحين بالأخلاق الكريمة ، لتساهموا فى بناء مصر الخالدة فى ظل الحرية والمجد .

والله أكبر والعزة لمصر

القاهرة فى / سنة ١٩٥

رئيس مجلس الوزراء

Fig. 1.5 The reply from president Abdel Nasser to young Zewail, 1956.

Fig. 1.6   A photo of Ahmed Zewail joining secondary school, 1957.

also participated in the school's athletic activities as well as practiced hunting for birds using rifles.

Ahmed then moved on to attend high school. His daily routine was about an hour of walk from home to school. That was how far our home was from school. Ahmed's tendencies towards scientific subjects became more apparent. He was fond of solving mathematical problems in mathematics, physics and mechanics. He felt the pleasure of victory each time he was able to solve a difficult question. He also helped his classmates to solve the problems in the classroom and explain to them the detailed workings. What intrigued my brother most were how things work.

He did not like the conservative ways of learning but preferred to absorb. Since then his mind was always glowing with questions: "Why?", "How?". And he did not just read the theoretical part, but one day decided to do a practical experiment in his room — to do a test, on his own, on how to turn a small piece of wood from solid to

Fig. 1.7   Zewail (second from right) holding a rifle with his hunting team at school.

gas. He succeeded in seeing the transformation process, his very first scientific experience, but the success almost turned into a tragic fire. He was saved — thank God. In high school, he also learned photography. In his final year of high school, Ahmed felt the pressure, but he was able to achieve high grades in chemistry, physics and mathematics.

1.3 Heading to Alexandria University

My father was not keen on Ahmed joining either Cairo or Alexandria University, as they were too far. It would require long trips and other hardships. He suggested Ahmed to choose the Agricultural Institute in Kafr El-Sheikh near Dessouque, where he could go on to become an agricultural engineer after graduation. Fortunately, our mother, who was religious and not good at either reading or writing, like most of the mothers at the time, interfered and supported her only son to realize his wishes to join the university in either Cairo or Alexandria. Ahmed exercised a few options and waited for the decision of the Organization Office of University Education.

A few weeks later, the postman delivered the letter with "Ahmed Zewail, student in the Faculty of Science, Alexandria University" typed on the envelope. Ahmed's childhood dream and that of his family became a reality. Ahmed then made the big trip to Alexandria University.

Life in Alexandria was more modern than in Dessouque, but that did not distract Ahmed. He studied hard and concentrated deeply to get distinctions.

Fig. 1.8   Ahmed Zewail with his colleagues in Alexandria University.

Ahmed graduated from the Faculty of Science, Alexandria University in the summer of 1967. It was a difficult time in his life and for the whole nation. June 5 was the beginning of the Six-Day War. Together with his friends, he had to wear a military uniform and volunteer in the civil defence. The day of his graduation was filled with mixed emotions for the whole family — there was happiness and also frustration.

The general sadness that dominated Egypt at that time did not prevent us from celebrating Ahmed's distinction in chemistry with honors. He was awarded the BSc. in Chemistry with the grade Excellent with Honors. He was appointed a demonstrator ("Moeid") at the Chemistry Department of the Faculty of Science of Alexandria University. For the first time, he printed visiting cards with his name and affiliation. After several years, he was awarded the Masters Degree in Chemistry in the field of Solar and Renewable Energy.

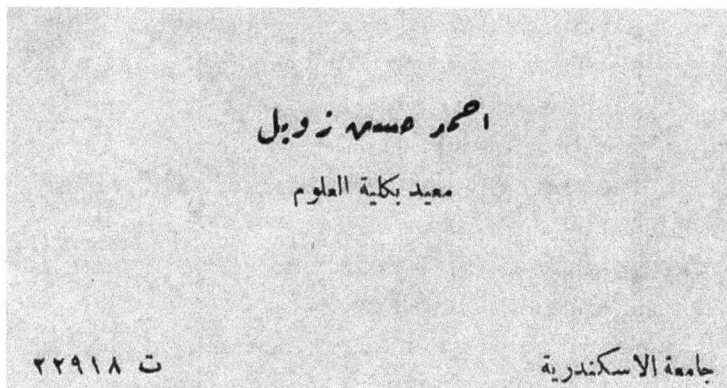

أحمد حسه زويل

معيد بكلية العلوم

جامعة الاسكندرية   ت ٢٢٩١٨

Fig. 1.9   Ahmed Zewail's visiting card written in Arabic after his appointment as demonstrator at the Faculty of Science, Alexandria University.

1.4 Experience of Ph.D. Studies in the USA

At the beginning of, 1969 he wrote to a number of American universities looking for a PhD Grant in Chemistry. That was the start of his ambition to go to the United States of America, hoping for another great journey and a desire to complete his PhD there.

After three months, the postman came again, carrying an envelope written in English, "University of Pennsylvania, United States of America". Ahmed opened the envelope and started to read the letter eagerly while praying to God with wishes to fulfill his dreams. He did not believe his eyes and read the letter more than ten times while thanking God at the same time. He was granted a full scholarship, which also included tuition and accommodation.

He was all ready to travel to the United States, the semester was to start August the same year. Despite knowing his departure would make his whole family sad, especially his mother, who was attached to him since he was the only son, the feeling of joy and pleasure somehow made everything alright. Ahmed tried to console my mother and ease her grief by telling her that his studies there was only for four years whereas it was six years. I was amazed at the role everyone had played in realizing my brother's dream, from our family to the small community of Dessouque. In this respect, we owe them lots of thanks. My brother boarded the plane and looked out of the window to catch a glimpse of the last few moments before the plane took off. All the effort to get the necessary approvals, certificates, signatures and seals between the offices and departments of the Ministry of Higher Education at Cairo and the Faculty of science at Alexandria University had come to this.

My brother was touched by the farewell scenes at the airport. Our mother's warm tears and farewell words had deeply touched his heart, who felt an unbearable pain thinking of the thousand miles separating a son from his mother, family, friends and homeland. However, he did not forget for a moment how blessed he was and prayed to God for his assistance. He quickly gathered his strength.

He was visibly worried, especially if you take into account he was going to a foreign country with only $40 in his pocket (that was the maximum amount allowed for every local passenger leaving Egypt at that time). Although the future was vague with an unknown pathway in front of him, his only focus was science and how he could reach perfection. My brother Ahmed did not surrender to his fears and the spirit of optimism was very strong within him. He had complete trust that God would help him and had faith in our mother's prayers. The years of difficulties he encountered in Egypt were the best training that would help him overcome difficulties overseas. My brother arrived in Philadelphia Airport on August 24, 1967.

In the States, he felt a feeling of helplessness like in a vast ocean. But he had no choice but to practice floating to avoid drowning. To survive there, he needed to overcome four major obstacles:

1. He needed to master the English language.
2. He needed to be aware of the strained relationship between Egypt and America at that time.
3. He needed to adapt to the difference in culture and lifestyle there.

4. He needed to learn quickly the complex laboratory equipment necessary for his experiments.

When Ahmed started to be deeply involved in the scientific studies, he found that Western scholars had neglected the Arab contributions to science that flourished during the Islamic Golden Age from the eighth century to the eleventh century. This disregard made him realize the importance of studying history (a topic he did not like in school). Reading history became his favorite pastime, especially the history of science, from which he learned about the decisive role played by the Arab scientists during the intellectual and scientific movements which led to the Arab Renaissance, as well as readings into the history of nations which had succeeded in building their future.

Ahmed succeeded in crossing the parameters of the vast ocean of scientific challenges and achieving the dream of his childhood and was awarded a PhD in Chemistry. At the young age of 27 years old, he was known as Dr. Ahmed Zewail.

1.5 Post-Doctoral Research

It was then the time to make an important decision, either to return to Alexandria University or to join the largest laboratories in America to receive a postdoctoral fellowship. The obvious choice was the latter and he submitted applications to five universities for a scholarship. He was selected by University of California, Berkeley for postdoctoral research in 1974. It was marked by a giant scientific projects funded by the US Government.

Two years later marked a turning point in his scientific journey when he arrived at the California Institute of Technology, "Caltech", with the consensus of the faculty. In 1976, he brought his family to the city of Pasadena. Being only 30 years old, it was not an easy task for him working at the university, especially with having senior scientists and Nobel Prize winners as co-workers. His longing to visit home intensified soon after. During the winter of 1980, he came back to his hometown for the first time after 11 years of separation from his family.

Fig. 1.10   Ahmed Zewail, at a young age of 27 years old, was awarded the PhD degree of Chemistry from the University of Pennsylvania, 1973.

Ahmed looked at us, eyes filled with tears of longing and nostalgia for his family, who he was not able to see except for the memories whispering in his ears carrying his mother's voice praying to God for him and her laughter as she prepares the peasant's food — Bram rice and the multilayered Egyptian pie.

He embraced us with so much warmth that one could never imagine it could be repeated one day. Ahmed felt the bitterness of the years of being away from his family, so he pledged to visit his family and homeland more often.

Fig. 1.11 Dr. Ahmed Zewail reunited with his mother after 11 years of absence, 1980.

1.6 The Road to the Nobel Prize

In the labs, Ahmed achieved great success in discovering the world of atoms at a close range and was even able to photograph molecules during their formation or dissociation that occurs within several femtoseconds. The time difference between combination and dissociation may be like the time difference between the 1 second and 32 million years.

Ahmed was just over 40 years old at that time when he made the phenomenal discovery that reveals one of the secrets of the greatness of Almighty the creator and the perfection of his creation. He was a single father at that time, after he and his wife — Mervat, the mother of his two daughters, Maha and Amani — were separated in 1979.

Fig. 1.12   Dr. Ahmed Zewail with me at Dessouque sporting club, 1980.

Fig. 1.13   Dr. Ahmed Zewail in his Lab. Caltech, USA, 1986.

Fig. 1.14   Maha and Amani, the daughters of Dr. Zewail.

My brother remarried his second wife Dima in 1990, who he had met in Riyadh during the the King Faisal Prize ceremony in 1989. They were blessed with two sons, Nabil and Hani Ahmed Zewail.

In May 1998, he was awarded one of the most prominent awards, the Benjamin Franklin Medal in Chemistry. It is one of the highest prizes of distinction in the United States, awarded only to the brightest scientists. It was sweeter for my brother as he was the first Egyptian-Arab Muslim to have been awarded the medal.

My brother returned home and celebrated this important accreditation with his friends and authorities in Alexandria and Damanhur.

The award had a great impact on the Egyptian community who would always be there to welcome him on each of his returns. They would converge on the streets surrounding Kafr El Sheikh's main square, where Zewail's statue stood together with the Governor, to celebrate Ahmed Zewail's outstanding achievements together with Zewail's family.

Fig. 1.15   Ahmed with his sons Nabil and Hani, 1997.

Fig. 1.16   Zewail raises his hand to greet his fellow countrymen in the celebration of Governorate of Behiera and Kafr El Sheikh.

Fig. 1.17   On the left: Zewail's statue at Kafr El Sheikh's main square with the Governor celebrating with the family. On the right: The Egyptian Government issued stamps in honour of Dr. Zewail — before and after Nobel Prize.

1.7 The Nobel Prize

On dawn of October 12, 1999, my brother received a phone call from the Secretary General of the Swedish Academy of Sciences. He was told that he had been awarded the Nobel Prize. After 12 years of research, and experimentation, he had won the world's biggest and most prestigious scientific honour.

The news was soon all over radio and television, but in his mind, he was thinking only of Egypt, his beloved mother and his family.

My brother returned to Cairo to celebrate the occasion with his family and also to receive the highest Egyptian honour — the Egyptian State Gold Nile Necklace — from the then President Mohamed Hosni Mubarak.

In 2005, in collaboration with his colleagues, he invented a four-dimensional electron microscope and patented it. In Ahmed's journey of science, what began as a fantasy and a question has now become a reality. However, the questions will never stop and will not end his journeys of science as knowledge can never be stopped.

Fig. 1.18   Ahmed with our mother visiting the pyramids after his great success.

My brother became the leader of a whole fleet of researchers and scientists on a continuous and distinguished scientific journey to wider horizons of scientific achievements. His personal dream had led him to an enriching journey.

In his dream, he has included Egypt, the Arab world and everyone else in the world who needed to embark on a journey towards a better future. He called for the development of education as a basis for economic development and the advancement of the lives of citizens and he hoped to achieve all this through his initiative — the Zewail City of Science and Technology.

His visit to Egypt in 2007 was his last meeting with his mother — may God have mercy on her. It was her wish to see my brother one last time. She got her wish and she passed away soon after.

Fig. 1.19   Ahmed Zewail receiving the Nobel Prize from the Swedish King.

Fig. 1.20   Ahmed Zewail receiving the highest medal "Egyptian State Gold Nile Necklace" from President Mohamed Hosni Mubarak, President of the Republic of Egypt.

In 2009, my brother was chosen to be a scientific adviser to the White House during Barack Obama's Administration. In the same year, he became an envoy to the Middle East to promote international cooperation in the scientific fields. Ahmed was able to achieve all this through three key principles — faith, clarity of vision and hard work. He was well aware of the importance of time and he was always full of enthusiasm and optimism. As a man of knowledge he was confident not only in building himself up, but also in building bridges between him and others, so that they too can succeed.

In 2012, he saw his dream became a reality with the official opening of Zewail City of Science and Technology (ZCST). The opening of ZCST has placed Egypt on the global scientific map. With an area of 200 acres and with the support of the Egyptian authorities this giant project provided the platform for distinguished scientific figures, Nobel

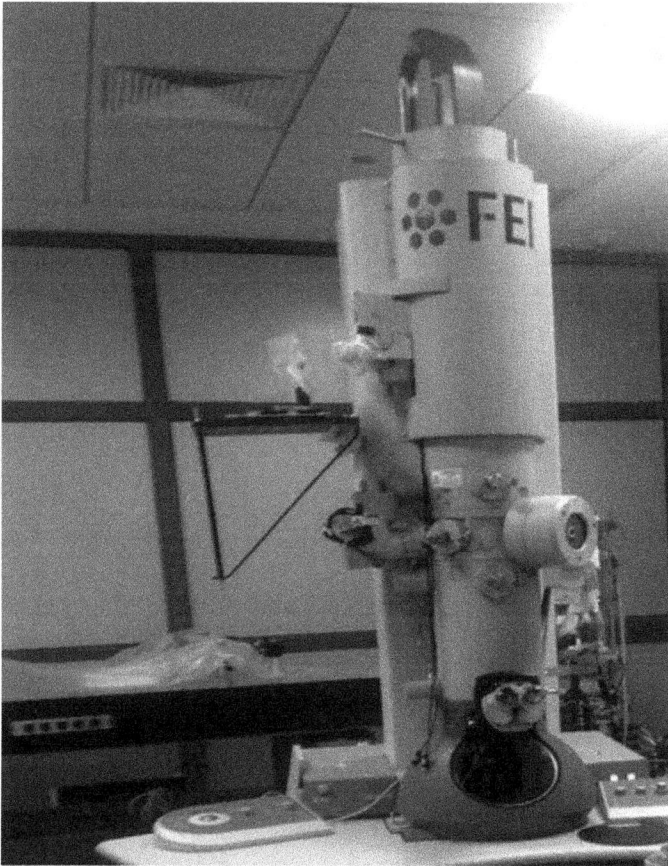

Fig. 1.21   The 4D Microscope in Zewail City for Science and Technology.

laureates and other specialists from abroad to cooperate with Zewail towards the success of this project.

Ahmed donated the entire value of his Nobel Prize towards the ZCST project and also collected billions of donations from a number of public figures and others who loved and trusted him. The first batch of ZCST students, who got the highest grades in high school, were exempted from paying fess due to their scientific excellence. In 2016, he donated the 4-D electron microscope to ZCST.

Fig. 1.22   Last meeting of Ahmed with his parents.

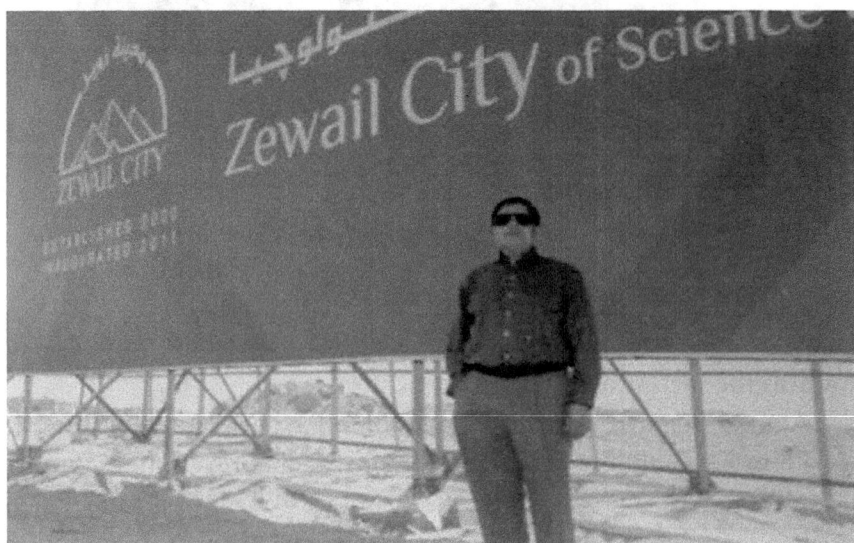

Fig. 1.23   Zewail in front of the Zewail City of Science and Technology.

1.8 Farewell Talk

On 26 February, 2016, which incidentally was Ahmed's 70th birthday, Caltech decided to celebrate the passage of 40 years of his presence at the university.

The ceremony was attended by a group of scientists who were also Nobel laureates. My brother gave his thank-you speech while standing between the flag of Egypt and that of the USA. He also insisted on putting the image of Egyptian President Abdel Fattah al-Sisi next to the image of Obama, the former American president.

Fig. 1.24   Ahmed Zewail at Caltech celebrating his 70th birthday, 2016.

Some of the attendees expressed their concern for my brother's well-being during his speech due to the intense dazzling lights during the ceremony as he was just recovering from his "illness".

On 2 August 2016, the journey of this extraordinary man came to an end. I received a phone call on that Tuesday evening telling me of God's will. The news did not hurt me alone, but all who had loved my brother from all over the world.

In the month of Ramadan in 2016, slightly less than two months before his death, his son, Nabil, told him that I was sick. He called me immediately late at night from the USA. He said, "I cannot sleep before I hear your voice and do not say to me that you're tired." For the first

Fig. 1.25

Fig. 1.26   Military funeral for Egyptian Nobel Prize winning scientist Ahmed Zewail.

time, I cried aloud. We talked for about 55 minutes. He said, "Do not cry, you are strong. I will come to see you after Ramadan and pass the Eid with you. . ." After that I had a feeling it would be the last call between us. My brother's long journey of giving his all and persevering out of love for his country and family in Egypt has ended. He finally returned, but only to have his body received at Cairo airport to be buried in Egypt.

This is the text of his will:

"Wash me by the Nile water before it is lost, wrap me in a cotton cloth of this kind land, bury me in its ground in which my tired bones will be relieved.

Illness and migration in the lands of knowledge hurt me. I paid my days to raise the name of Egypt between the nations; I used my knowledge to defeat illness and spent my days like a monk in laboratories . . .illness won and succeeded. . . but my cresset will always illuminate and shine, on top of my grave will grow an evergreen plant. . . Remember me whenever a new bud grows. . .remember and pray for me."

Dr Ahmed Zewail

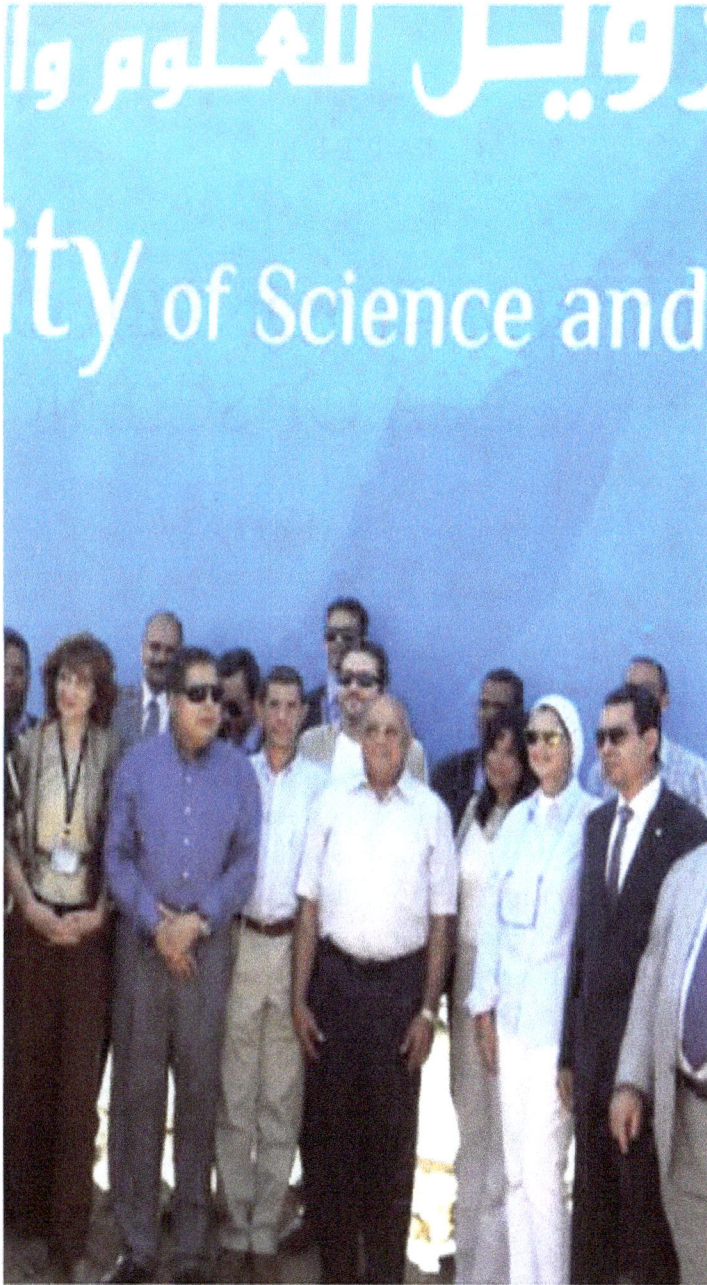

Fig. 2.1

CHAPTER 2

———

ZEWAIL — PIONEER OF A NEW ERA OF SCIENCE

Mahmoud Abdel-Aty

I t began in 1997, when I was a Ph.D. student at the Max-Planck
Institute of Quantum Optics (MPQ) in Munich, Germany. At that
time, there was an annual visit to the institute by a group of top
scientists from all over the world. They were there as external moderators
to review the research that had been done in the institute; that was when
I first met Professor Ahmed H. Zewail, looking very authoritative among
the scientists and brimming with enthusiasm. It was my first meeting
with him, but definitely not the last. I could never have imagined that
I would go on to develop a strong relationship with him and his dream,
which he was always talking about in his scientific meetings, symposia,
and cultural interactions with the many people from within and outside
the academia. Zewail was not only a maestro of knowledge in chemistry
and physics, but also a very charismatic character and this was very
obvious to anyone who dealt with him. Listening to his talks during his
stay at the institute made me proud to be an Egyptian. His talks were
riveting. He would always give encouraging comments on the work
and efforts of others. We remained in contact via e-mail and met at

different international conferences and exchanged knowledge and had chats together.

As a dreamer, he was thinking of emulating what Max Planck had achieved in harnessing human powers to build a prosperous place and applying the same concepts to Egypt. Zewail had the idea of merging the strengths of the Egyptians with the knowledge and wisdom of Egyptian scholars. He was always talking about Egypt, especially in terms of the current state of science in Egypt. He would like very much to have a small piece of Max-Planck planted in Egypt that works in a similar manner to Max-Planck and even Caltech, that is, by bringing eminent Egyptians from all over the world to such a set-up and providing them with a conducive and supportive environment, so that their efforts can be converted into technological and economic progress within Egypt. A portrait of Ahmed Zewail — a prominent African — would be hung at the entrance of the proposed academy together with portraits of other Nobel Prize winners from Africa, and I would be keen to deliver this

Fig. 2.2   Max-Plank Institute for Quantum Optics, 1997.

message to him, "The whole of Africa is really proud of you!" This is the dream.

Zewail City of Science and Technology (ZCST) is the science hub that he dreamt of establishing in his lifetime after he received the Nobel Prize. This had always been in his mind — the founding of a science center at the forefront of the world's development and the cutting-edge in science, and he never gave up that dream. Now, Zewail's City, located at the October Gardens near the pyramids in Giza, is operating as the center-of-excellence in many diverse fields of science, engineering, and technology. One may have noticed the correlation between the pyramids, Ahmed's dreams, and Zewail City's logo.

Only outstanding students are admitted to study there and brilliant researchers and scholars are being recruited to do research and to

Fig. 2.3   Author Mahmoud Abdel-Aty posing with students of the ZCST.

Fig. 2.4   Students of ZCST, 2014.

teach the frontiers of their respective fields. Zewail City offers top-notch programmes at the undergraduate level, from engineering to science and technology. The students are enthusiastic about joining the prestigious hub of learning that Ahmed Zewail had dreamt of and built for 13 years.

I joined Zewail City in 2014 after spending many years at Bahrain University, first as a Professor of Mathematics and later as the Chairman of the Scientific Publishing Center of the same university. I remembered before I left Bahrain University, I visited Zewail City and witnessed the fantastic progress of the researchers and academicians there. Fortuitously after the visit, I received an invitation to head the Mathematics Department — a crucial component of Zewail City's academic syllabus. It was an honor to leave my position at Bahrain University for Zewail City — and return to Egypt carrying with me the optimistic spirit to contribute towards the building of a new Egypt together with the rest of the academic community in Zewail City.

My appointment at Zewail City coincided with my daughter's application to Zewail City as a student for her bachelor degree in 2013, and her subsequent enrolment with a full scholarship in renewable energy engineering. She and other bright students liker her are the new face of Egypt — one that is marked with knowledge and excellence — and this is a return back in history to what Egypt was well known for.

Many meetings were held among the various communities in Zewail City during which Professor Zewail was careful to inculcate the spirit of hope in all of us. I can still feel the family environment, and remembered very well when he first held his meeting with the first batch of students: he shed tears at seeing their achievements and, at the same time,

Fig. 2.5   Tasnim Abdel-Aty was awarded the bachelor degree in renewable energy engineering program of ZCST in 2014.

Fig. 2.6   Zyad, a second year student, posed a brilliant question to Dr Zewail, 2015.

achieving his long-time dream and witnessing the continuation of his legacy.

Ahmed Zewail was a sincere person, and he was genuinely interested in building the cultural and scientific characters of the students within Zewail City's campus. I could still remember when I attended a scientific lecture in 2015 at Zewail City by Zewail's postdoctoral fellows. Zewail was in attendance — sitting in the first row — when Zyad, a second-year student, asked a brilliant question to one of the postdoctoral fellows. The question went something like this "…but what about molecular vibrations of such corresponding nanosystems…what if, if, and if …?" Zewail insisted on replying, took the microphone and gave an answer to the question, which came from someone sitting at the back. Zewail knew Zyad, but didn't recognize the voice at first. Zewail then asked, "So what's your name?" Zyad replied, "Hey, I'm Zyad from the back." Zewail then turned to see Zyad, and he complimented Zyad by saying:

"Oh! It's Zyad — Zyad who is very much following our work and he even understands details of my work better than I do!"

This is just one example of how Ahmed Zewail was a source of support and inspiration. He has left a huge impact on the people whom he had worked with; in the academia, governments, agencies, and many other places. His fingerprints on the progress of the world, including the economic development resulting from the applications of his achievements, will remain forever. On behalf of all of us who knew this honorable man, Ahmed Zewail, we will miss him dearly.

May he rest in peace.

Fig. 3.1

37

CHAPTER 3

HOW DID I GET TO KNOW HIM?

Farouk El-Baz

A s the Apollo missions to the Moon began in 1969, NASA thought of establishing a Lunar Science Institute (LPI) close to its Manned Spacecraft Center in Houston, Texas. Scientists involved in the program, from within and outside of NASA, would meet annually in March to discuss the results of studying and interpreting the data collected by the astronauts. A decade after the Apollo missions had ended, we continued to meet and discuss research results. During a break at the 1982 LPI meeting, Gerald (Gerry) Wasserburg, a noted geochemist and expert on age-dating of lunar rocks at the California Institute of Technology (Caltech) approached me to ask:

"Do you know Ahmed Zewail?"
"No, I don't," was my answer.
"Well, he is a young compatriot of yours at Caltech."
"In what field?" I asked.

"Chemical reactions and he is an up and coming guy."
"Great!" I said. "Say hello to him for me."

I was reminded of that conversation four years later during a discussion with another Caltech colleague, as relayed below.

After the Apollo Program ended in 1972, NASA asked me to head the Earth Observations and Photography Experiment on the Apollo-Soyuz Test Project (ASTP) that was planned for July 1975. It was the first joint American-Soviet mission to orbit the Earth. While planning activities for that mission, I was invited in 1974 by President Anwar Sadat to start studying the deserts of Egypt in support of his visionary program: "Greening of the Desert." I began field work with colleagues from Ain Shams University and the Egyptian Geological Survey. The work resulted in planning sites for photography and observation by the astronauts.

One of my ASTP team members, Leon (Lee) Silver, a noted structural geologist who became President of the Geological Society of America, was a Caltech Professor. Knowing of the result of my worldwide desert research, he wished to invite me to lecture on the topic. In 1986, he approached me to ask:

"Have you met your countryman Ahmed Zewail?"
"No, not yet," I said, and added, "but Gerry Wasserburg told me about him."
"Well I wish you would come lecture about your desert research at Caltech."
He said, "This would be a good opportunity to meet him."
"Great, that's a deal," I agreed.

I felt very proud to hear two of America's most noted academics speak of a fellow Egyptian scientist. I prepared slides to show what we had learned about the deserts of Egypt in addition to those in the rest of North Africa, the Arabian Peninsula, Australia, India, China, and to the American Southwest. With my set of slides, I flew to Los Angeles, where Caltech is, to give my lecture. What I did not know was that Ahmed came to the lecture hall, and sat quietly at the back.

At the end of the discussion period, he came over to introduce himself to me. We then walked over to his laboratory, where he showed me his research equipment and introduced me to some of his colleagues. It was a great meeting that made me even more proud of him and his accomplishments.

Then, came the wonderful news of his selection for the Nobel Prize. It was the first to an Arab scientist, and only the second to someone from Islamic country, after M. Abdus Salam of Pakistan. Salam was a good friend of mine who had established The World Academy of Sciences (TWAS) in Trieste, Italy, to which I was elected in 1985. I recommended Ahmed, too.

My next meeting with Ahmed was in a US-Egypt political setting. In 1997, President Bill Clinton invited President Hosni Mubarak for talks in Washington, DC. President Clinton further came up with the idea of a joint meeting of the two of them with a group representing Egyptian professionals in the US. Thus, the White House invited ten notable Egyptians for a round table discussion. As we met in a reception room before entering the meeting place, Ahmed came in and stared shaking hands and introducing himself. When he approached me, he retorted, "Finally, someone I know!"

In 2002, Ahmed and I were to work together when both of us served as members of the Board of Trustees of the Library of Alexandria. According to the governing laws of the Library, the President of Egypt heads its Board of Trustees. However, President Mubarak did not wish to do so and asked Mrs. Mubarak to serve in his place. It was indeed a very good choice as she did a wonderful job in dealing with the international board members as well as the library personnel. It was during those encounters that Ahmed would tell me about his wish to establish a scientific research organization in Egypt for nurturing young Egyptian talents.

"There must be many like me and you," he would say.
"They need us to help them along," he would add.

Naturally, I was in full agreement with him, promising to help in that endeavor in any way I could. His dedication to the idea made it possible to imagine a first-rate scientific research operation — on par with his group at Caltech — to be managed fully by Egyptians for the benefit of scientific progress in Egypt. What could be a more noble cause than that?

In 2007, Ahmed began to discuss his initiative with Egypt's Minister of Scientific Research. His plan was that the initiative (Zewail City of Science and Technology) would benefit future generations of Egyptian researchers. I would follow up with my words of support. The Government appeared to have approved of the idea in technicality, but questioned the preferred location. He called me one day and he was fuming,

"The Minister changed the location without informing me!"
"Take it easy, that can be corrected," I said.
"Location is something that can be negotiated with him," I added.
"I will not talk to him," he said emphatically, "You call him!"

I accepted the responsibility and promised to call the Minister on an upcoming trip. As it turned out, the Minister was changed not long after, and the tense situation eased. As things began to improve, Ahmed started thinking ahead by establishing a Board of Directors. He invited me to join the Board, but I had to decline and apologized to him because I was already over committed. I told him my cardiologist had advised me to drastically cut down my international forays.

"I do not need you to travel," he said.
"Great, so there is no need for me to join the Board," I added.
"No, I need you on the Board to call you before or after the meetings."

"But, I might also need you to talk to somebody in the Government," he added.

That was a good arrangement, and I already had indicated I would help in any way I could without having to add to my over-committed schedule. Also, his choice for Board members included people I already knew well and highly respected, which included Prof. Lotfia El-Nadi, Egypt's foremost nuclear physicist with great international stature, H. E. Mona Zulficar, one of the best legal minds in the country, and Dr. Fathi Saoud, a friend and colleague of mine from Ain Shams University, and former President of the Qatar Foundation. Thus, both me and Ahmed agreed I would be a member of the Board, but would not be required to attend its meetings.

Things moved smoothly and Ahmed's dream became a reality when the Egyptian Government assigned it to the location of Nile University — an initiative of former Prime Minister Ahmed Nazif. That selection did cause problems with students of the said university, and Ahmed called me again for help. I communicated with several of the students as I had met some at the Arab Stars of Science competition.

Fig. 3.2   Dr. Farouk El-Baz in the Arab Stars of Science competition.

I also met with the group that had camped out in protest near the start of the Cairo-Alexandria road. It was very clear the assignment of the location of the site was not done with great care or with much thought by Government agencies. However, it was later revealed a proper location was finally selected for the Zewail City to satisfy all the needs, and to avoid any conflicts with existing organizations.

Dr. Ahmed Zewail left us too soon. Let us pray to God to bless his soul. His passing was duly mourned by the whole nation in a parade befitting a great hero. He was a man of principle who worked tirelessly for the cause of knowledge. His many accomplishments filled us all with pride as we looked back at his distinguished career. The most enduring part of his legacy was his untiring effort towards the establishment of a first-rate research institution in his native land. His wish was to create an environment that would encourage future generations of Egyptian scientists. His dream was for some of them to excel in the search of new knowledge for the benefit of humanity. Indeed, may God bless his soul.

Fig. 4.1

CHAPTER 4

THE SCIENTIFIC FINGERPRINTS OF AHMED ZEWAIL IN THE ARAB REPUBLIC OF EGYPT

Lotfia El-Nadi

In the Name of Allah, the Most Beneficent, the Most Merciful. This chapter is about one of Egypt's most outstanding individuals, who devoted his life to perfecting the teaching of scientific knowledge. He did this by utilizing his God-given abilities and, in the process of doing so, made the groundbreaking scientific discovery of how chemical molecules are formed and how they disintegrate. He discovered this by imaging the dynamics of chemical bonding at the femtosecond scale. Such an innovative idea does not just flash in the mind of any ordinary person. It can only happen to someone who has insight into, and a deep scientific understanding of the laws of nature.

Dr. Zewail was one such person — inquisitive and knowledgeable. I hope to give readers some idea of the genius of this scientist and how he helped Egypt and Egyptians improve in the fields of scientific education and research. Ever since he gave his first lecture at the American University in Cairo in 1989 to announce a new branch of science — femtochemistry — Dr. Zewail would share scientific ideas with me personally whenever we met.

Zewail had built what could be described as the world's fastest camera, with a shutter speed of just a few femtoseconds. Zewail's "camera" is actually a combination of instruments which includes an ultrafast laser that emits flashes of light in femtosecond pulses, better known as a "pump beam", and a spectrometer that detects the presence of different atoms and molecules by measuring the spectrum of light they give off when hit by the laser beam in a destructive process called "dissociation". Another beam of laser light travelling a few femtoseconds after the first one, known as a "probe pulse", hits the molecular pieces — which are flying apart at that moment — at different intervals, consequently revealing how long it takes for various chemical elements to appear and in what order they appear. The results of this landmark experiment, performed by Zewail and his students, can be observed in a molecular movie. In that movie, hydrogen iodide (HI) and carbon dioxide (CO_2) played the leading role. The recording is phenomenal not only because it's filmed at the femtosecond scale, but also because it showed evidence of a brand new scientific discovery, i.e., the kinetics of the formation, dissociation, and disintegration of chemical compounds. This would not have been possible before Zewail's "camera" was invented. Hence, Zewail was awarded the Nobel Prize in 1999 for his invention of the "camera" and for the scientific discovery he subsequently made.

Prophet Mohamed (the peace and blessings of Allah be upon him) said in the saheeh hadeeth that, "The scholars are the heirs of the prophets." (As related by Tirmidhi, Abu Dawud, Nasa'i, Ibn Maja, Ahmad, Ibn Hibban, and many others.)

"O God, have mercy on Ahmed Zewail and keep him in your paradise with the good, the righteous, the faithful, and the good companions."

Appended below is a list of some of Zewail's fingerprints in Egypt as represented by his contributions as an individual in rendering assistance to the universities and institutions during his visits to Egypt. At Cairo University in particular, we have seen and recorded the aspirations and achievements of Egypt, and we seek to fulfil the former and build upon the latter, following Zewail's 2011 slogan:

"EGYPT CAN"

4.1 First Meeting with Dr. Zewail at Evart Hall of the American University in Cairo (AUC)

In the last quarter of 1989, I received an urgent telephone call from my colleague and dear friend, Dr. Salah Arafa, Professor of Materials at the Chemistry Department of the American University in Cairo, in which he mentioned Dr Ahmed Zewail's visit and invited me to attend his lecture on his recent discovery.

I had read his latest research, which was published in prestigious journals, e.g. two of his papers were published in *Science*. These were early Zewail publications about his ingenious invention. The numbers appearing on the left denote those on his list of publications:

Real-Time Femtochemistry
A. H. Zewail and R. B. Bernstein
Kagaku to Kogyo (Science and Industry) 41, 298 (1988)

Femtosecond Real-Time Observation of Wave Packet Oscillations (Resonance) in Dissociation Reactions
T. S. Rose, M. J. Rosker and A. H. Zewail
J. Chem. Phys. **88**, 6672 (1988)

Real-Time Laser Femtochemistry: Viewing the Transition States from Reagents to Products
A. H. Zewail and R. B. Bernstein
Chem. Eng. News, **66**, 24 (1988)

Femtosecond Clocking of the Chemical Bond
M. J. Rosker, M. Dantus and A. H. Zewail
Science **241**, 1200 (1988)

Femtosecond Real-Time Probing of Reactions. I. The Technique
M. J. Rosker, M. Dantus and A. H. Zewail
J. Chem. Phys. **89**, 6113 (1988)

195. Femtosecond Real-Time Probing of Reactions. II. The Dissociation Reaction of ICN
M. Dantus, M. J. Rosker and A. H. Zewail
J. Chem. Phys. **89**, 6128 (1988)196

198. Laser Femtochemistry
A. H. Zewail
Science **242**, 1645 (1988)

I was 55 years old at that time. Dr. Zewail was only 43 years old, but he had already published nearly 200 scientific papers. I had heard about Dr. Zewail before when he held a scientific conference at the Institute of Graduate Studies, University of Alexandria, which was headed by Professor Sadr, one of Egypt's finest scientists at that time.

I decided to attend in order to learn more about his research. I said to myself that "this is a golden opportunity that should not be lost". Perhaps groundwork could be laid for cooperation between us and the young American-Egyptian scientist, whose research was a proof of his ability to discover and innovate, especially since at that time I was starting a collaboration with the University of Texas in the United States through the American Society of Optics. I had also submitted a proposal for a major project, i.e. the establishment of an Institute of Laser Sciences at the Faculty of Science to start laser research in all areas that serve the community — industry, medicine, environment and communications — and I was also working towards the formation of a scientific team of researchers in laser science, to be funded by US AID with a non-refundable grant.

My first meeting with the world-renowned Egyptian-American scientist, Ahmed Zewail, at the Evart Hall of the American University in Cairo in late 1989, was an amazing experience. He first explained how photographic resolution has improved 10 billion times since the 1887 sequence (shown below) of a horse galloping 10 meters in 1 second, as compared to a sequence that showed hydrogen iodide colliding with carbon dioxide — creating carbon monoxide, hydroxide and iodine — which was captured during one trillionth of a second.

Fig. 4.2   Sequence of a galloping horse.

He was a master of presentation and was able to explain his idea clearly. His listeners could easily comprehend the scientific details of his idea and responded by asking questions. Zewail answered all the questions, even questions that were trivial, with confidence and with an open heart, always with a welcoming smile on his face. He was an angel sent by God to enlighten our minds. My hope at that moment was that we could win him over for Egypt.

I had the honor of talking with him at tea time, when I explained to him what research and projects I was carrying out and the difficulties we were confronting. I can still remember his comments, "You, Dr., have an extremely long gut yourself, Dr...." We had a good laugh when we talked about what managing projects and waiting for approvals were like in our country.

I invited him to visit the laser research centre at the Physics Department of Cairo University to give a lecture to the researchers.

This was initiated by me with the backing of the Academy of Science and the KFA at Jülich in Germany.

Fig. 4.3 Zewail lecturing to researchers at the Laser Center, Physics Department, Faculty of Science, Cairo University, 1989.

Zewail was very enthusiastic about the invitation. He gave a wonderful lecture that garnered such huge interest that many could not be accommodated in the lecture hall. We were obliged to put up speakers so that his lecture could be projected to those outside the hall. Admiring the audience and their questions, Dr. Ahmed promised to come back soon.

He started by sending us his contact details. Our President invited him to Cairo University, and he accepted the invitation. He promised to stop by on his way back to the USA after receiving the King Faisal Prize.

4.2 First General Lecture in Cairo at the Ceremony Hall of Cairo University

This was the title of Dr. Zewail's first lecture at the University of Cairo:

"Egypt's Civilization 6000 Years Ago: An Imaging of Flemish-Based Plasma Particles Formed and Analyzed by Dr. Ahmed Zewail, A Pioneer in Their Discovery and in Research on Them."

The hall was full of Egyptian scientists such as Professor Mohamed Abdel Maqsoud El-Nadi, internationally-renowned Professor of High Energy Physics; Professor Nayel Barakat, who was collaborating with Imperial College London on fiber optics research; and other members of the Faculty of Science, Engineering and Medicine. Many attendees came from the Egyptian universities: the Military Technical College, the National Research Center, the Atomic Energy Commission, and the Center for Laser Research and Applications. For the first time, Dr. Dima Al-Faham, the young bride of Dr. Ahmed Zewail, attended a lecture on his discoveries in his great country, Egypt.

He accurately explained his brilliant ideas in utilizing femtosecond lasers to determine molecular dynamics, elucidating the facts that were hard to grasp for others.

He introduced his research topic with a motivated determination that shone light upon his findings and displayed his polished skills

Fig. 4.4   Some of the attendees of Dr. Zewail's first public lecture at the Great Ceremony Hall of Cairo University in 1991.

Fig. 4.5   Lotfia El-Nadi introduces Ahmed Zewail from Cairo University on the platform. From the right: Professors Ahmed Samy, Abd El Shakour Dean of the Faculty of Sciences; Najib Helaly Gowhar, Vice President for Graduate Studies and Research; Ahmed Zewail; Lotfia El-Nadi.

Fig. 4.6   On the left side Zewail is being awarded the First Golden Medal of the Topical Society of Laser Sciences (TSLS) and on the right side he is being awarded the Shield of CAIRO University.

in communicating. He spoke confidently in front of the audience in a way that permitted everyone to understand and follow his ideas. With a smooth and pleasant smile, Zewail not only explained his scientific ideas, but also his plans to build up his country, Egypt, which had housed a great civilization for more than 6000 years. Floods of questions were asked. As I realised then, his smooth and convincing answers downplayed the greatness of his discovery while disclosing its rich secrets. In this way, he cemented his reputation as a modest and honest scientist.

The Hall exploded with the sound of thousands of clapping hands. I had never witnessed such a reception before.

He was then awarded the first Golden Medal of the Topical society of Laser Sciences (TSLS), followed by the Shield of Cairo University, which is usually only awarded to the kings and presidents of nations.

After this meeting, we celebrated with a number of friends and members of Egyptian society; including Dr. Adel Imam, the famous Egyptian doctor, who said to Dr. Zewail that he was an extraordinary phenomenon in the history of science. In fact, he became exactly that.

4.3 Cooperation with Professor Dr. Ahmed Zewail

As Dr. Zewail was a close friend, I often consulted him about the facilities of the National Institute of Laser Sciences. I even asked him to help me choose the name of the institute in English, since the abbreviation of the name of the institute would be NILS if we left it as it was. He thought hard while we were having dinner together and said, "Call it the National Institute of Laser Enhanced Sciences — NILES." What sheer brilliance!

Soon he left for Caltech, and I was appointed Chairman of the Physics Department in the Faculty of Science of Cairo University for the third time.

I asked him to enter into a bilateral agreement with Caltech, whereby he would become one of the founders of NILES. He answered me, "Lotfia, you have been working on establishing NILES since 1985. That is nearly nine years. How can you ask me to be one of the founders?" He was adamant in refusing to take credit for someone else's effort. He suggested that we write a personal agreement that would allow my students to be trained in his laboratories at Caltech. He would give them a scholarship and my project would pay their travel expenses, so I wrote an agreement and signed it with him. This document is still with me. He asked me to visit his lab in Caltech whenever I can.

Dr. Zewail's visits, follow-up discussions, and encouragement continued to benefit me and the entire project team. He was so involved in our projects that he accompanied us in inspecting the construction of the institute and gave us new books, so that we could follow the latest developments in laser physics and chemistry.

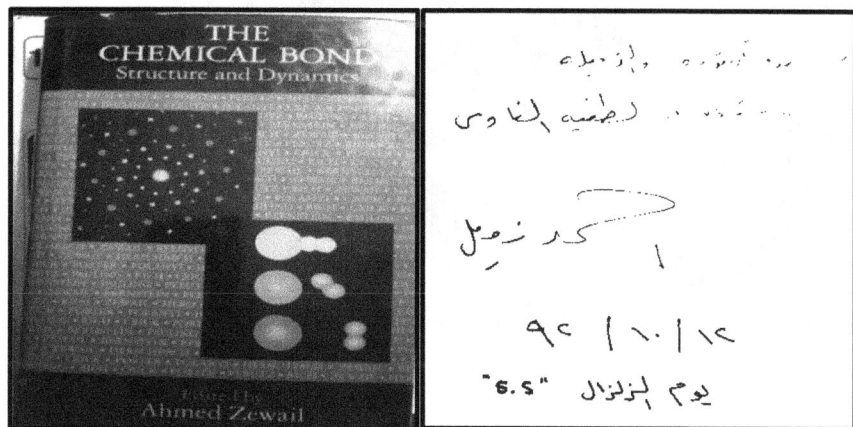

Fig. 4.7   Professor Zewail donated this book to me during the 12th Octobeer 1992 earthquake in Cairo.

I traveled to Los Angeles in 1993 to attend the American Optics Association (OSA) conference. Zewail received me at the airport and introduced me to the scientific community at Caltech. I had already seen his laboratories during a short previous visit in 1992. At that time, lots of students were astounded by his laboratory, as can be seen in the picture. He greatly honored me by insisting that Caltech host me during this visit. He invited me to his home in Pasadena and celebrated my visit with his family. He also showed me his study at home, which was often filled with the enchanting songs of Om Kalthoum and other Egyptian musicians. He was so proud of his nationality.

I then left to attend the OSA conference at LA and, to my surprise, I found Dr. Zewail there. He did not tell me that he would be present.

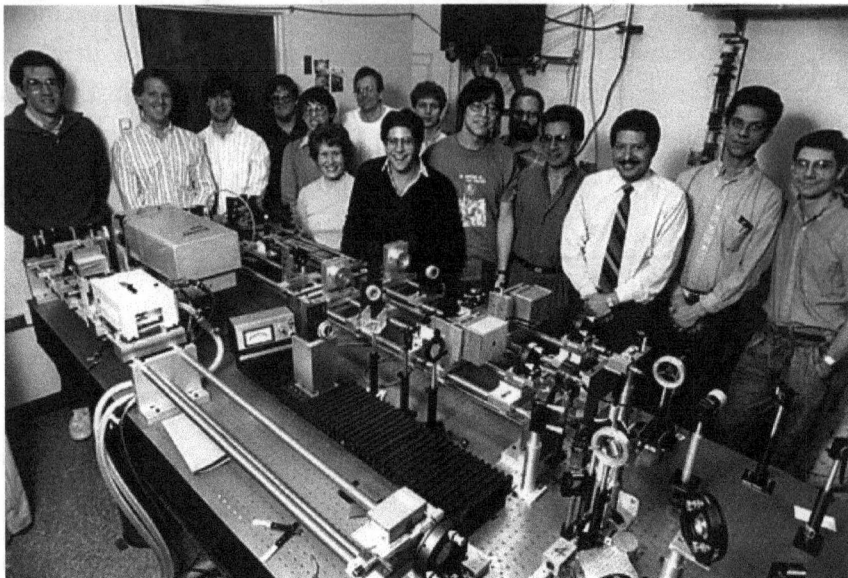

Fig. 4.8   Professor Ahmed Zewail with his students and colleagues in his laboratory at Caltech. On display is the equipment for the femotosecond laser which he developed. This being the first experiment to use a dye laser, he was keen to adjust the ingredients with his own hands.

Dr. Zewail was there among the other international scientists. Everyone was racing to catch a moment to talk to him and learn about his research. I attended his presentation and, for the first time, realized the importance of Dr. Zewail.

4.4 The First International Conference on Laser Science and the Opening of NILES at Cairo University

As mentioned previously, the scientific meetings between Dr. Zewail and our project team continued through e-mails and him personally whenever he came to visit Egypt. Professor Zewail had promised to accept the honorary chair of the first international conference, which would be held at the opening of NILES in March 1994. He accepted the invitation despite being busy with his research.

The National Institute of Laser Enhanced Sciences, or "NILES", had been progressively established through the great efforts of a team of my students, who worked day and night to put the finishing touches to the labs and workshops. We were racing against time to accomplish nearly everything.

On the following page, one may find information about "NILES".

Dr. Zewail informed me through e-mail that he had already invited around 40 scientists from USA, Europe, Canada and many other countries throughout the world. He told me they would come free of charge, without claiming travel costs to and from their country, not because they loved the conference, but because they wanted to meet Dr. Zewail. God bless him.

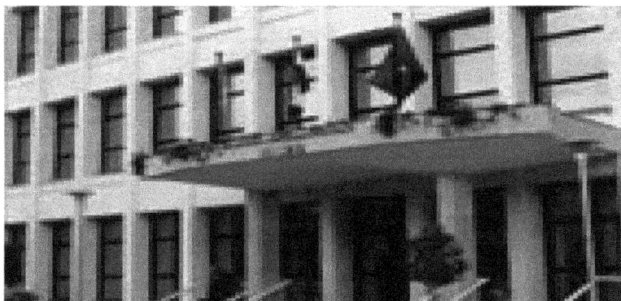

NILES

Fig. 4.9 The National Institute of Laser Enhanced Sciences (NILES) was opened in 1994, at Cairo University.

NILES was initiated as a project by Professor Lotfia El-Nadi in 1985 and started in 1990 as a Laser Center run by the Faculty of Science. Under her direction, the opening of the National Institute of Laser Enhanced Sciences (NILES) was undertaken in March 1994. The First International Conference Honorary President was Professor Ahmed Zewail and the conference was chaired by Professor Lotfia El-Nadi. NILES became an independent institute in 1995.

OBJECTIVES

➤ Training a generation of Egyptians to have a high level of scientific and technical knowledge in the laser-enhanced sciences.
➤ Developing laser sciences in most fields of application.

GOALS

➤ Introducing cutting-edge technology into several fields.
➤ Catching up with laser studies as carried out by the international scientific community.

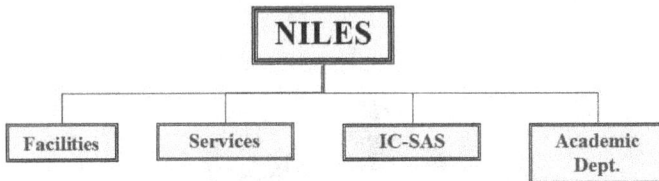

NILES is the only center of excellence in Egypt & Africa in the field of the laser-enhanced sciences & their applications.

Fig. 4.10   120 ps/2J neodymium glass laser and target chamber.

The first session of the First Conference of Laser Science was well-attended with guests coming from all over the world to meet with Dr. Zewail at the sidelines of this important event. Amongst those attending were Prof. Bloembergen, who won the Nobel Prize in 1993; Dr. Margret and Mornan and her husband, Dr. Kaiptyn of the University of California, currently one of the most important scientists working on attosecond lasers; Professor Svanberg of Uppsala University in Sweden and Dr. Catherine, his wife, a laser cancer specialist from Stockholm Hospital in Sweden and one of the world's most famous medical scientists. Many professors from Germany, France, England, Italy, Spain, India and Canada, who are pioneers in their specialties, can also be seen in this photo.

In addition, the Egyptian scientists are present, God bless them all, as well as some young researchers and members of the project. The conference was a success that would not be easy to match. The opening speech was given by Dr. Zewail, who was awarded the Highest Shield of Cairo University by Prof. Nageeb Hosny, president of Cairo University.

Fig. 4.11 Some participants of the First Session of the First Conference of Laser Science.

Fig. 4.12   The representative of the Prime Minister, Prof. Nageeb Hosny, president of Cairo University, and the Dean of the Faculty of Science, Prof. Samy Abdelshakour.

Fig. 4.13   Dr. Zewail being awarded the Shield of Cairo University.

During repeated scientific meetings with me and the researchers at the institute, which were held until the end of 1994, Dr. Zewail donated a large number of scientific articles to me. He also gave me the annual Science Encyclopedia, which contained, at the time, an important introduction to genomics and predictions about the future of research in this important scientific field.

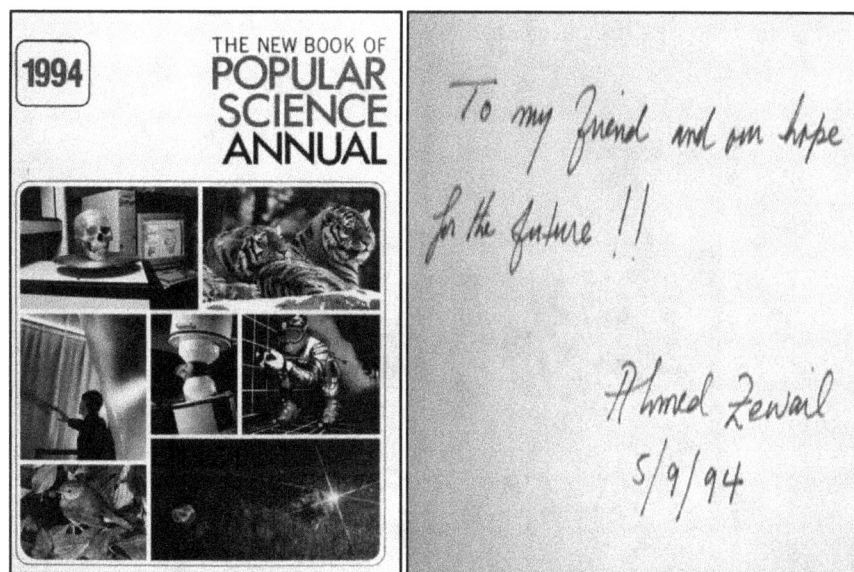

Fig. 4.14   The New Book Of Popular Science, Annual Edition — a book about the importance of science — which was donated to me by Professor Zewail. He wrote a message in it describing me as "our hope for the future".

Zewail was appointed honorary member of the Topical Society for Laser Sciences (TSLS) in early 1992. His constant commitment to advancing the limits of science and technology for the benefit of mankind, and his legacy of scientific discoveries will remain a source of inspiration for scientists. We believe that he will never be absent from our hearts. Indeed, Eman Raslan, one of our famous journalists, described him as: "a sun that will never set".

Dr. Zewail donated the following periodicals, which were of great help to us, to the library of the TSLS: Chemical Physics Letters, Nature, and Science.

From 1993 to 2003, the library of the Topical Society of Laser Sciences (TSLS), had a total of more than 1500 periodicals which were internationally renowned. The presence of these periodicals in the TSLS library encouraged the community of Egyptian researchers to publish their scientific findings in them.

Fig. 4.15   Some of the cabinets we are currently preparing to fulfill our commitment to Prof. Zewail to publish in science, particularly in the field of laser science.

4.5 Dr. Zewail Encouraging Attendance at International Conferences

Zewail was keen to keep us abreast of the dates and places of international conferences and urged us to consider the necessity and importance of participating in them.

4.5.1 Femtochemistry: The International Conference of Femtochemistry, Ultra Fast Chemical and Physical Process in Molecular Systems. (Lausanne, Switzerland, 4–8 September 1995)

In 1995, I was invited to participate in the Third International Conference on Femtochemistry, held in Switzerland, at the University of Lausanne, with Majed Chergui as President of the Conference. Dr. Zewail was the one behind this invitation.

The conference was attended by a great community of laser scientists. My paper was accepted for presentation as a poster. It was then published in the conference book without any expenses being incurred and this was, of course, in response to Professor Zewail's encouragement. It was one of the most successful conferences I had attended. Professor Zewail's presentation was entitled: "Femtochemistry — Advances Over a Decade".

I still receive questions from scientists about my publication in the proceedings, which is:

W. Tawifk, M. N. Omar, Y. Gamal and L. Nadi (1995). Bulk and Surface Effects in Liquids due to the Interaction of High Powered Pulsed Laser Beams. *Proceedings of the International Conference of Femtochemistry, Ultra Fast Chemical and Physical Processes in Molecular Systems*, 483.

The article can also be found as the 56th article in my list of publications (http/:www.lotfianadi.name.eg).

4.5.2 Fifth International Conference of Femtochemistry and Biology (Toledo, Spain, 2–6 September 2011)

Dr. Ahmed Zewail was awarded the Nobel Prize in December 1999. He was made the President of the Fifth International Conference on Femtochemistry and Biology, which was held in Toledo, Spain, in September 2001. He invited me to attend the conference. It was at that time that I had to come back to my laboratory and do an experiment to visualize biological cells using a femtosecond laser. I accepted the

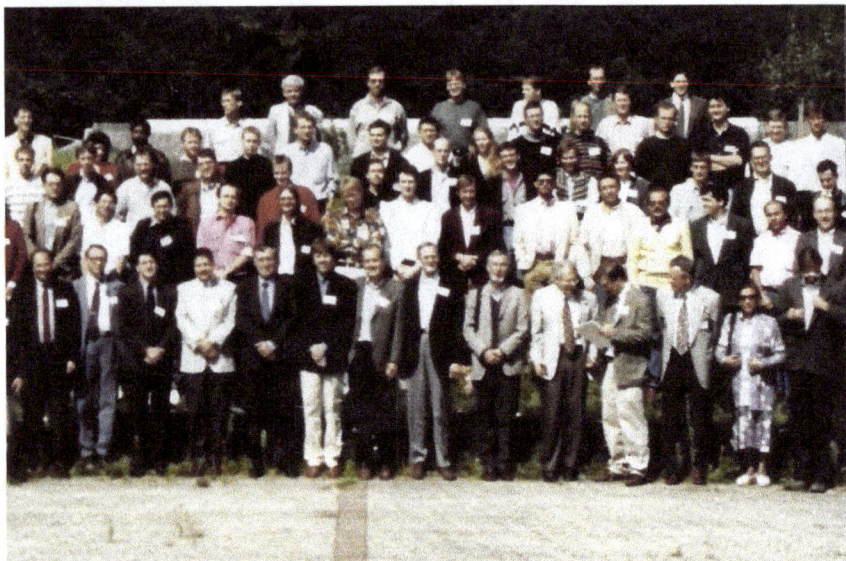

Fig. 4.16

FEMTOCHEMISTRY – ADVANCES OVER A DECADE

AHMED H. ZEWAIL
Arthur Amos Noyes Laboratory of Chemical Physics
California Institute of Technology
Pasadena, California 91125, U.S.A.

1. Introduction

The Lausanne Conference, organized graciously and masterfully by Professor Majed Chergui, has continued the tradition of the series on *femtochemistry*, in which the first meeting was held in Berlin (1993), the birthplace of molecular dynamics in the group of Michael Polanyi. This year's conference featured more than 250 contributions from 25 nations. Two highlights that we all remember are the exciting and diverse presentations of lectures and posters, and the collegial and supportive discussions throughout the entire conference. We are grateful to Professor Chergui for all his efforts and for the hospitality in wonderful Lausanne!

Femtochemistry experiments performed on calculators and computers go back six decades when H. Eyring and M. Polanyi (1931) developed the first potential energy surface and J. Hirschfelder, Eyring, and B. Topley (1936) did the first trajectory calculations with femtosecond steps. With femtosecond time resolution, it became possible to observe and study molecular dynamics in real time. Over the past decade, the field has witnessed an explosion in the range of applications [See, e.g., the books referenced 1-6] and in the studies of molecular dynamics in different phases [7, 8]. This article, based on the lecture given at the Conference, focuses on the progress of research at Caltech. After discussing a few examples of elementary reactions, particular emphasis is made on applications to complex molecules and reactions, including isolated systems, clusters, and the condensed phase. As shown in Figure 1, these applications are only a fraction of those discussed at the conference by many colleagues.

Before discussing recent progress in the studies of molecular reaction dynamics with examples from organic, inorganic, and physical chemistry, we shall give a brief description of the concepts involved and reflect on the key elements of the approach.

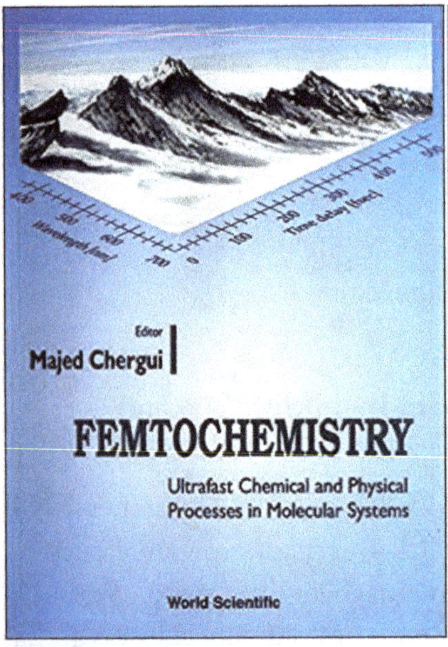

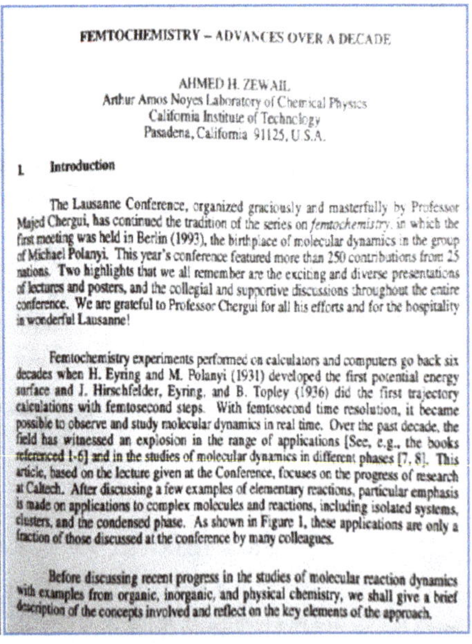

Fig. 4.17 The cover of the proceedings followed by the introduction of Dr. Zewail's presentation.

invitation so that I could consult Dr. Zewail about my important findings. I was always keen to follow his scientific views — how could I not be! He held a very prestigious position in the hall of science.

I presented my work orally. It was met with lots of questions and comments, and it was published in the proceedings of the conference as:

L. Nadi (2002). Ultra Fast Lasers to Determine Molecular Biological Structures. *Conference Proceedings, Femtochemistry V,* 255.

It is also available as the 36th article on my list of publications (http/:www.lotfianadi.name.eg).

Fig. 4.18   The Fifth International Conference on Femtochemistry and Biology. From right to left: Dr. Mostafa El-Sayed, Georgia Institute of Technology; Dr. Zewail; Dr. Martan, Head of Chemistry Department at the Ecole Normale Supérieure in Paris; Dr. Santa-Maria, Dean of the Faculty of Sciences, Climensky University, Toledo; Dr. Lotfia El-Nadi; Dr. Abdelrazzak Duhal, Head of the Femto-Laser Laboratories, Climensky University, Toledo.

Fig. 4.19   From right to left: Dr. Abdelrazzak Duhal, Dr. Ahmed Zewail, Dr. Santa-Maria, Dr. Magid Chergui.

4.6 First International Conference on Modern Trends in Physics Research at Cairo University

The First International Conference on Modern Trends in Physics Research was held from 4th to 9th April, 2004, under the Honorary Chairmanship of Nobel Laureate Ahmed Zewail at the Grand Hall of the University of Cairo. It was opened by the Minister of Higher Education and Scientific Research and the President of the Cairo University.

Dr. Zewail's honorary chairmaship and his Nobel Laureate status persuaded a great number of distinguished professors from around the world to attend the conference. The attendees were mainly working on the modern disciplines of physics. They came from more than 35 countries, representing organizations from 15 countries, in addition to the Egyptian and Arabian universities and institutes.

The first words were spoken by Dr. Zewail, who "praised him and his attendance at the conference". Everyone was surprised when he announced for the first time his invention of electronic photography in four dimensions. The instrument he had invented is now known as the 4-dimensional electron microscope. We realized that what he had just presented constituted a breakthrough in physics research.

Fig. 4.20   Shown in the picture: Professor Ahmed Zewail, Honorary President of the fourth Modern Trend in Physics Research; President of Cairo University; Dean of Faculty of Science Cairo University; Minister of Higher Education; Professor Lotfia El-Nadi, Chairman of the Conference.

Fig. 4.21

At the end of his talk, the Ceremony Hall exploded with the tremendous sound of clapping hands. Finalizing the opening session, the President of the University awarded Professor Zewail the Conference Shield as is clear in the photo below.

Fig. 4.22   Zewail being awarded the Conference Shield from the President of the University of Cairo.

4.7 Fourth International Conference on Modern Trends in Physics Research, December 2010

The Fourth International Conference was held under the Honorary Presidency of Dr. Zewail from 12 to 16 December 2010 at the Faculty of Science of Cairo University and Sharm El-Sheikh. After fulfilling his duty to invite distinguished professors in the fields of femtosecond and high-powered lasers, he apologized because his duties had increased after being appointed as a scientific adviser to President Obama in the USA.

Fig. 4.23

He wrote some introductory words, which were placed on the front page of the proceedings of the conference, and was published by World Scientific.

AHMED H. ZEWAIL

NOBEL LAUREATE
HONORARY CHAIR MTPR-010

The International Conference of "Modern Trends in Physics Research" MTPR-010 is the Fourth of the Bi-annual Int. Meetings organized in Cairo, Egypt by the Physics Department of the Faculty of Science, Cairo University. It was my honor to agree to be the Honorary Chairman of this Conference in response to Prof. Lotfia El Nadi's invitation, but due to my obligations I was unable to attend, but would follow up its progress.

The conference had participants from around the world and many were from Egypt. The conference had attracted well-known scientists in their fields offering their experience and exchange ideas with researchers from Egypt.

The success of this conference is due to Professor Lotfia El Nadi, Program Chair of MTPR-010, efforts, devotion and determination as well as her colleagues who were involved with the organization and logistics.

I personally would like to take the opportunity to thank also World Scientific Publishing for publishing this proceedings and I look forward to future MTPR's.

Ahmed H. Zewail
Caltech, Pasadena, California, USA
Zewail City for Science & Technology, Egypt

4.8 Zewail and Bibliotheca Alexandrina

Fig. 4.24

Fig. 4.25   Dr. Ahmed Zewail's public lecture at the Bibliotheca Alexandrina in April 2009.

The building of the Bibliotheca Alexandrina was completed in 2001 with the assistance of UNESCO. After its establishment, a Board of Trustees comprising the world's most prestigious personalities were appointed. Ahmed Zewail was chosen in 2002 as one of the personalities to support the scientific activity which the library aspired to. He was keen to attend periodic meetings of the Board of Trustees on a permanent basis. He was not satisfied with only attending meetings, but gave public lectures on the importance of broadcasting scientific thoughts to the youth. He also suggested serious solutions to Egypt's problems with scientific education and facilitated discussions at the grassroots level, which addressed the need to overcome the obstacles suffered by Egyptian society. He participated in many seminars and conferences held by the University of Alexandria and other universities at the Bibliotheca Alexandrina.

Zewail has accomplished numerous achievements and awards, some of which are listed below:

- Written or participated in writing about 600 essays and 16 books.
- A member of many international academies and societies, such as Academy I, Academy II and Academy III. [Pick three academies for examples, seven is too much.]
- A member of several boards of international institutions and some of the most prestigious companies.
- He has been a visiting professor of excellence at a number of universities around the world, delivered close to 500 keynote speeches, and attended many university graduation ceremonies and public lectures.

"In the world today, we depend on each other more than ever before in the history of humanity," Zewail wrote in his book *Voyage through Time: Walks of Life to the Nobel Prize*, in which he expressed his hope for bringing together civilizations and promoting peace and human dignity. He continued, "And I hope that the new century and the new millennium will have many achievements for both cultures, which will bring about the renaissance of Egypt and provide a new perspective for humanity as a whole. As I have said many times, I am full of optimism."

His Excellency has recommended that all of Ahmed Zewail's cultural and social books be donated to the Bibliotheca Alexandrina to benefit all the sons of Egypt.

4.9 The Origin of the Idea for Establishing the Zewail City of Science and Technology

Dr. Ahmed Zewail was awarded the Nobel Prize in Chemistry in 1999, making him the first Arab to receive the award in the field of science. He received the award for his pioneering research in the field of femtosecond science, which led to the founding of a new branch of physical chemistry. In the same year, Zewail was awarded the Order of Merit of the First Class and the Great Nile Necklace, the highest Egyptian medal. In 1998 and 1999, the postal stamps "Portraits" and "The Fourth Pyramid" were issued in recognition of Zewail's outstanding contributions to science.

In a meeting with the President of the Republic in 1999, Dr. Zewail expressed his desire to establish a scientific institution for science and technology. The President gave orders to Prime Minister Dr. Atef Ebeid to speed up the allocation of land in Sheikh Zayed City. Dr. Ahmed Zewail and Dr. Atef Ebeid laid the foundation stone of this institution in the presence of the Ministers of Higher Education and Scientific Research, Housing, and Communications at that time.

Fig. 4.26 The 1999 celebration of the laying down of the foundation stone of the scientific institution of Zewail City for Science and Technology at the official site of the ZCST in the city of Sheikh Zayed.

In 2011, following the revolution of 25 January, Zewail City of Science and Technology was inaugurated. A dream that Zewail had had since 1999 — to be its president — came true. Zewail had always aspired to launch a scientific renaissance in Egypt.

The Supreme Scientific Advisory Council was formed on December 21, 2011 and was comprised of distinguished scientific figures, as shown

Fig. 4.27   The Supreme Advisory Council in front of the Garden City housing, 2011. From the right: Magdy Yaqoub, Professor Ahmed Okasha, Professor Ahmed Zewail, Lady Mona Ezz Eddin, Professor Mohammed Abu Al-Ghar, Professor Fathy Saud, Dr. Ahmed Galal, Professor Mostafa El Sayed and Professor Lotfia El-Nadi.

Fig. 4.28   Last meeting of the Boards of Trustees and Consultants of Zewail City of Science and Technology 2015-2016. From the right are the professors Mohamed Ghoneim, Lotfia El-Nadi, Ragai Atieh, Ahmed Okasha, Amr Salama, Yoshar Abbas, Ahmed Zewail, Nabil Elaraby, Ayman Abu Hadid, Mona Zulfikar, Mohamed Abou El-Ghar, and Mostafa El-Sayed.

in the picture that I received by e-mail from Mrs. Ragia Mansour, Executive Director of the Board of Trustees of Zewail City.

There were many factors that led to the later changes to the boards of trustees and consultants. The meeting shown below was one of the last councils under the chairmanship of Prof. Dr. Ahmed Zewail. It happened in late 2015.

Dr. Ahmed Zewail explained the policies to be followed with regard to some of the obstacles that necessitated the moving of the project to the new headquarters.

However, Dr. Ahmed Zewail was taken from us by God before the completion of this new site. God willing, he will support us in spirit. With the sponsorship we have received, we can surely complete the project.

Fig. 4.29   The new campus of Zewail City for Science and Technology.

"O God, have mercy on Ahmed Zewail. Allow him your paradise with the righteous believers, and the faithful and good companions."

Fig. 5.1

CHAPTER 5

MY BIGGEST LOSS — AHMED ZEWAIL

Mostafa A. El-Sayed

After getting a Bachelor's degree in the US academic system, students interested in getting a PhD apply to different schools with doctoral degree programs. Those schools accept the best students who applied. The schools having good research professors, who usually select the top students that applied. The number of applicants they accept depends on how many teaching laboratory courses they offered.

Good research schools usually accept large numbers of these graduate students. The new students are taught laboratory courses in their first year. In the second year, continuing until graduation, they get funds from the research grants of the faculty member they select

---

Opening photo description: During a scientific meeting on Spectroscopy attended by (from right to left) — Dr. Ahmed Zewail (far right), Professor Robin Hochstrasser (with whom Dr. Zewail studied for his PhD at the university of Pennsylvania), Professor Andy Albrecht, (Cornell University), Professor Mostafa El-Sayed (University of California at Los Angeles) and Professor Thomas Spiro (Princeton University).

to work with. The graduate students get their doctoral degree for the completion of their research.

Before coming to the United States, Ahmed sent me a letter asking if I would accept him in my research group if he was admitted to the University of California, Los Angeles (UCLA). That year was my sabbatical year; after teaching for five years, the faculty in good universities usually has the right to take a year off and go away to do research with other professors. The same year, I was invited by the American University of Beirut and I accepted as I wanted to take my wife and children to see my family in Beirut. While in Beirut, I could invite my family in Cairo to come and meet me and my family in Beirut, since at that time I could not go to Cairo and leave at will.

My secretary replied to Zewail without showing me the letter, since I was already in Beirut. She told him to apply to our university (UCLA) and if he was to be accepted in Fall, I could discuss with him about doing research and joining my research group. Ahmed kept this letter with him as he thought it was I, who had told her to do that, to get rid of him.

He went on to apply to do his PhD at the University of Pennsylvania and got accepted by Professor Hochstrasser, who was doing research, which was very close to my own at that time. Professor Hochstrasser was a good friend of mine. Being in the same research area, Professor Hochstrasser and his group went to the same scientific meetings as I did. In one of the meetings, Ahmed showed me the letter and suggested I did not want him in my group. I never knew about him applying to join my group until that day. My secretary never told me about it. That was how I met Ahmed and got to know that he had applied to join my group. Ahmed went on to become a professor at California Institute of Technology (Caltech).

We became good friends and I was a sounding board for Ahmed. I gave him advice and wrote him letters of recommendation. It was very easy for me to see how bright, highly motivated and hardworking he was. He wanted to stay in the States. After receiving his PhD, I wrote a letter supporting his intention. Since we were both in the same scientific field, it was natural for me to be his supporter as well as advisor in his scientific career.

It was easy to write strong recommendation letters for him as he showed excellent talent in research and in giving excellent talks. When he wanted to do his postdoctoral research, he sought my advice as to who he should work with. I suggested doing postdoctoral research with a friend of mine (C. Harris) at the University of California, Berkeley.

After completion of his postdoctoral research, I told him to teach at Caltech. The university contacted me by phone to get my support. UCLA (where I was working) and Caltech are within 40–50 miles from each other and previously, I was a postdoctoral fellow at that fine institution. The faculty there and I were very good friends. They saw my strong letter of recommendation for Ahmed and contacted him. After consulting me, they made him an offer.

Ahmed and I were very close. We met once a month at an Arabic restaurant with Arabic and stage entertainment to talk about the research done by our groups. I used these meetings to learn about Ahmed's research and wrote letters in support of him to keep his research well exposed. I was the Editor of the Journal of Physical Chemistry then, so I invited him to write feature articles frequently to expose his exciting research to the US scientific community.

I invited him to give talks at every national meeting I organized. He was very talented in giving talks and wrote brilliant scientific articles. The university where he taught, Caltech, is a highly respectable research institution and that made it easier for him to show his scientific talents, attracting brilliant young students and postdoctoral fellows. His brilliance in research led to his rapid ascension up the scientific ladder. For every scientific meeting that me or my colleagues organized in our respective fields we made sure to invite Ahmed so that he could present his excellent research.

It was always a pleasure to meet and sit in the same room with Ahmed, talking about the scientific future of Egypt, at the various national and international meetings. We discussed the lack of research funds in Egypt and also the importance of research in developing the industry that can help the Egyptian economy.

It was clear in our discussions that a place like Zewail City was needed. In our discussions, Ahmed had an additional idea for research based on Egypt's natural resources. This part of the City will be in charge of the logistics such as Egypt's natural resources, the amount of resources and the associated cost in development of products. Unfortunately, the idea was scrapped due to limitations in funding, but hopefully its formation will only be delayed until the economy is stabilized.

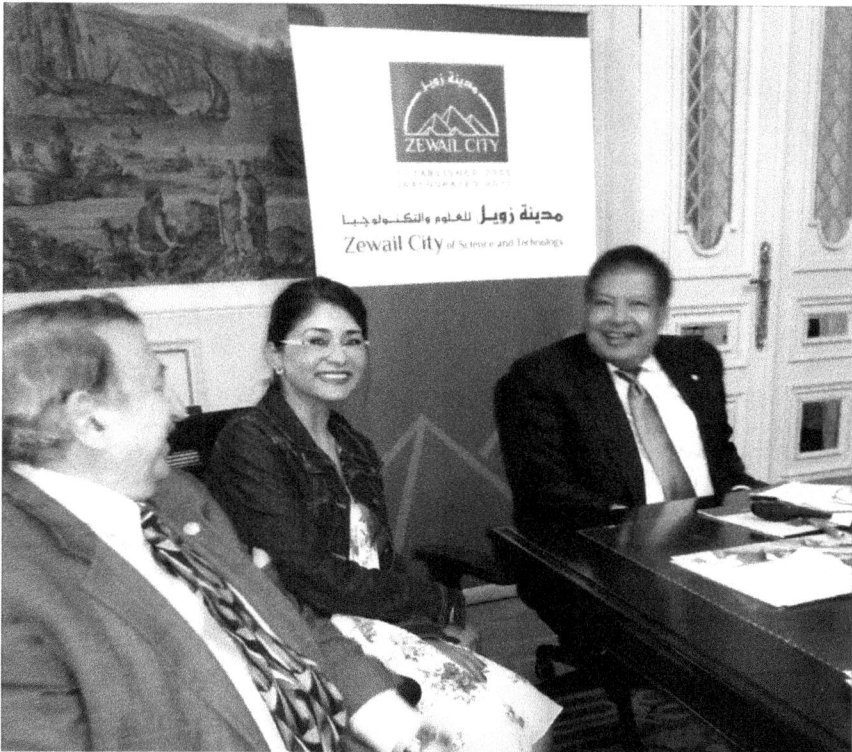

Fig. 5.2 The author, Mostafa A. El-Sayed, with Ahmed Zewail in between the official board meetings in 2015. They were joined by the author's granddaughter to discuss scientific activities with Zewail.

Ahmed was sharp in identifying any economic developments and trends. In science, he was equally as sharp — perhaps more — in predicting technical directions. He was able to predict the outcome of many scientific changes and results before doing the experiment. While doing research, he was able to keep his sense of humor, at least in the early stages when we crossed paths a great deal. We hope that Zewail City will continue to help Egypt toward the future that it deserves, the one that Zewail has worked hard for during his lifetime.

Fig. 6.1

81

CHAPTER 6

MY EXPERIENCE WITH DR ZEWAIL

Mohamed Th. Hassan

As the days of our life pass, only a few remarkable moments shape our life and last in our memories forever. They lurk in the shadows of our minds, bringing back all the times that will be forever unforgettable.

Deep in my heart, my journey with Dr. Zewail will always be the happiest yet the saddest memory. More than a year has passed since he left us, yet he is still here in every move. Five years ago I met him and since then he has changed my future and my whole life.

Every moment spent with this glorious scientist is historic. He has proven fulfillment, patriotism, and dedication to his work that will keep him a milestone in the history of a nation. He has carved out his own path in the history of mankind. His story will be told over generations, holding one and an ongoing title "Dr. Ahmed Zewail". In the coming lines, there are some one-off moments that portray my experience with Dr. Zewail — a quick look at the fascinating world of one of the brilliant minds of the century.

It was almost mid July 2012, at Switzerland, when I first met Dr. Zewail. It was during one of the most important conferences in the field of ultrafast sciences. On the night before my talk, while I was preparing my presentation, I thought of sending an email to Dr. Zewail to invite him to attend my talk. He was one of the keynote speakers of this conference, and he was due to give his lecture at noon, almost three hours after my lecture. Although I knew that it would be difficult for him to come because he is usually extremely busy, I sent him the email anyway.

The next morning, I proceeded to give my talk, which had very good feedback from the audience. Dr. Zewail didn't show up, as was expected. During the break time after the first session ended, there were many questions from the attendees and our discussion continued throughout the break. On my way back to the conference session, I saw Dr. Zewail walking next to someone whom I learnt later was currently a professor at École Polytechnique Fédérale de Lausanne (EPFL), the university that was hosting the conference, and who was one of Dr. Zewail's disciples. When he saw me, he yelled, "Mohammed! Mohammed!" Then he asked, "Did you send me the email earlier?" I answered, "Yes, I am [sic]". He then invited me for coffee and he asked me to tell him more about my work. Although he admired my work very much, he asked in a wondering tone, "How, in the first place, did you make it to Max Planck and join Ferenc Krausz group?" He knew that this research group at Max Planck is one of the best in this field worldwide.

I said, "I have tried a lot and got rejected five times before I was finally accepted in this group and I have done my best to prove myself." He further asked me how I had managed to publish all these papers in *Nature* and *Science*. I replied, "Since I have joined them, I have been working very hard to catch up and rise to their notable level."

He smiled with satisfaction and asked me about my future plans. I said, "I have learned much in Germany, but I think it would be better now to move to the US." I expressed my intention to move either to the group of Prof. Leoni, whom I had already spoken to him, at Berkeley

University or Prof. Bakusbaum at Stanford University. He looked at me and said, "Well, send me your CV and your published papers." I replied "Ok" without asking for any reason. After spending two hours talking, it was the time for his lecture. He went to the stage and I took my seat among the audience to listen to him. That was my first meeting with one of the most iconic scientists in the world. I cannot express my sincere gratitude for the exceptional time I had in his company.

After returning to Germany, I was surprised when my professor Dr. Ferenc called me to his office. He asked me, "Why didn't you tell me that you want to leave the group and join Dr. Zewail's at Caltech?" "But I didn't ask him", I replied and I told him about my meeting with Dr. Zewail at the conference. He was stunned, adding, "Dr. Zewail had phoned me and asked me to issue a recommendation letter for you to be able to carry on your hiring procedures at his research group at Caltech." He was very upset, but he knew that I was saying the truth. He told me that he couldn't let me go now, saying "I am going to call Dr. Zewail again to keep you working with me for another two years."

Right after I left Dr. Ferenc's office, I sent a message to Dr. Zewail asking him about what just happened. His secretary replied asking for my phone number. I sent it. Five minutes later, my phone rang. It was Dr. Zewail. He said, "Mohammed, you don't want to work with me?" I answered, "No, of course, it is an honor for anyone to work with you." He then asked, "Then why is Dr. Ferenc not allowing you to join my group?" I told him what had just happened with Dr. Ferenc. He humorously added, "He is my friend and if he doesn't send you immediately to join my group, I will not nominate him for the Nobel prize again, hahahahahaha." He then asked me to contact the secretary to complete my hiring process at Caltech. It was done in one week. I've started preparing myself for this next phase to move to US and to work with this extraordinary man. Actually, he didn't leave me much room of choice, he was known for always doing what he wanted no matter what it was. I felt that this was a destined step.

Fig. 6.2   A moment to treasure — discussing science with Dr. Zewail in front of Braun Laboratories.

From my first day with Dr. Zewail's group, it didn't occur to me we were both Egyptians. My intention was just that a student wanting to learn from his master. Soon after, we started working on our first research project. Despite already having a solid scientific lab experience at Max Planck, one of the greatest institutes in the world, there was always a new perspective that Dr. Zewail would share. His scientific know-hows were based on three main points. First, choose a new and unique idea that could possibly make a difference in real life, not one that is born and buried inside laboratories. Second, the results should be correct and accurate; and third, the research paper should be written in an elegant way. His experience and distinctive working style allowed him lead revolutionise science with foresight.

At that time, I had an idea to improve the speed of the microscope. He listened to me carefully and asked me to prepare a full presentation about how I would do it. After I had finished my presentation, he said that former members of the group have tried to work on this idea; however, they failed, so they stopped trying. He added, "You know, to do this, it will be a great challenge." I have been always been passionate to work on something that others think is impossible, so I accepted the challenge straight away. I got his approval and I believe he was proud of my choice. I knew that this was a step forward not only to my scientific skills, but to my personal relationship with Zewail as well.

Following Zewail's style, I started work immediately and once I have finished, I showed him the results. In a four-hour meeting, he scrutinised every single detail of the research and its results, and asked me every single aspect. He was very impressed and on the next day we sent the paper to be published. After only two weeks, the paper was published with a great recommendation from the reviewers. I could still remember the day when the paper was published. It was a very emotional moment — a moment of happiness and honor. Although, I had published many papers in *Nature* and *Science*, two of the top scientific magazines worldwide, this paper was very special to me.

I went to Dr. Zewail's office misty-eyed. Dr. Zewail's secretary noticed and asked me why, I replied, "Nothing, I think something wrong with my eye." Suddenly, I heard him coming out of his office and saying in Arabic, "We had always spoken in English so we could maintain a professional level of dealing," he laughed loudly and asked me, "Did the Germans make you stiff?" I said, "Sixteen years ago I joined the Faculty of Science to become like you one day, and today I am publishing a paper with our names side by side. That was a long journey." He gave me a hug and said, "One day you will be a great and renowned professor."

That was one of the happiest days in my life, a day that I won't ever forget. Since then, our relationship had changed for the better, we became closer. I regarded him more like a father. However, I maintained professional distance in our daily business interactions and he respected that.

Soon after, he asked me to his office and spoke to me about the "Zewail City" project, and the amazing effort he had put in to achieve this dream. This was not a personal dream; this was the dream of the whole nation — to build an institution with very high standards that offers a great opportunity for young Egyptians to learn and contribute towards the elevation of our great nation. He asked me to help him build the 4D microscopy lab and an imaging center at "Zewail City". Even if this would affect the pace of my research at Caltech, it was an honor and I agreed immediately. There were daily meetings to discuss all the details and updates related to "Zewail City". Despite his full schedule and his not so good health condition, he devoted time daily to follow up on the developments of the project. He was in constant contact, following up on the development of the construction of the new campus. At times, he would stay awake until very late, bearing more than he could, so he would be able to accomplish his dream.

One day, after our usual meeting and on his way out from the meeting room to his office, he fell at the lobby. I lent him my hand and helped him to sit on the nearest chair in his office. He didn't want to be seen in this situation, so I moved out quickly after I had informed the secretary to take care of him. He was working day and night nonstop. I always wondered how a man at his age and suffering from a serious disease could work like this, but this was Ahmed Zewail.

In early 2016, Caltech was preparing for a very important event — the 70th birthday celebration of Dr. Zewail. He was following up on all the ongoing preparations and gave instructions and advice to make sure that it would be perfect. He took great care and concern over the organization of the event, making an extra effort with his assistant and the secretarial staff as if it was his farewell party, his last.

The ceremony was scheduled on February 26th, 2016, and it was called "Science and Society". He invited some of his friends who had won the Nobel Prize in Physics, Chemistry, and Economics.

One week before the ceremony, he called the research group for a meeting to explain the program of the ceremony. He asked me to accompany the Egyptian media delegation in their tour of the 4D microscopy labs. I apologized to him saying I didn't like media appearances. However, I agreed on one condition, that my work be kept away from the cameras and that any of my colleagues would help with the tour of the labs. He laughed typically and said, "Ok and don't worry, you will be kept away from the lights." I knew he wasn't convinced with my statement. As expected, on the day of the ceremony, his secretary called me and asked me to take the Egyptian delegation for a tour of the labs. I was ready for such a surprise as I knew he wouldn't take no for an answer. After the tour ended at the microscopy labs, we headed to Dr. Zewail's office and that was when I saw him and noticed the superb effort this man was putting in. Although he looked very tired, he was fighting on strongly as usual. The media delegation did some short interviews with him before heading to the hall where the ceremony was taking place.

Everything was perfectly prepared and was going very well. In the first three hours, friends of Dr. Zewail gave their speeches and then it was Dr. Zewail's turn. He stepped up to the podium and started talking, looking visibly tired. His speech was not scientific as some had expected, but was more about his life since infancy and how his journey started back there at Damanhur until his arrival at Caltech with all the successes and failures, which he had experienced throughout this amazing journey. He spoke about the turning point of his life; winning the Nobel Prize and how it changed his whole life. Then came his dream, the building of Zewail City, how it was the most important thing in his life and all the related problems and obstacles that he had to face which almost ended his dream.

After 2011, the project was revived again. He showed photos from the City and some photos of the new campus. He was very cheerful about being able to present Zewail City to the world.

Fig. 6.3   A pose with the students before lecturing at the Zewail City.

Lastly, he mentioned about his illnesses and his fight with this merciless disease that he had suffered from and how he was able to overcome it with the help of his family and friends.

When he finished his lecture, all the attendees greeted him with appreciation. For a full ten minutes, they just stood there, expressing their gratitude and respect. He was touched and, for the first time, there was a glimpse of tears. At this very moment, you could see all the successes, illness, hopes, and dreams. It was a scene that resembled closer to a farewell to one of the world's greatest scientists of the 20th century.

In the evening, Dr. Zewail had a prolonged television interview with the Egyptian media. It was one of his best interviews. He opened his heart and talked about everything, even though he was very exhausted. During the break time, I whispered to him, "Great interview Dr. Ahmed, I know you are tired, may you get well soon." He said, "Mohammed, I just want to live to see the dream of Zewail City become true." At this point, I was even more worried about him and I believed that his words that day were like as if he was sending me a message that it would be his last to me.

During dinner, Dr. Zewail stepped up to the podium again, this time to give his closing speech, but it proved to be too much for him this time. He lost his balance and almost fell to the ground. However, like a moving scene you see in a drama or a movie, he managed to bring himself up with the support of the flag of Egypt that was placed near the podium. It was there for him, as he preferred it to be next to him rather than the American flag. He was Egyptian to the core and you could see it clearly in the eyes and you would always hear it in the unique accent of his village at Damanhur. He even asked to be buried in the land of his birth if he was to pass away.

After the ceremony, as was expected, there was so much talk about his health condition and whether it was getting worse. A few days later, he called for a meeting and, when I saw him entering the meeting room, I knew he wanted to tell us something important.

He started talking about his fight with cancer and about his recovery for the first time. He said he was diagnosed with cancer again and that he would have to go to the hospital to start his treatment.

Dr. Zewail was hospitalized for his treatment. I checked on him twice daily. He seemed ok and was getting better. In two months, he was back, after a surgery and some chemotherapy sessions. Although he was still very tired, he was there working nonstop as usual, both with his research group and following up on the construction of Zewail City.

He then asked me to go to Egypt to check on the ongoing projects there. A week before my trip to Egypt, Dr. Zewail didn't come to his office and I knew he was suffering from a severe chest cold. Yet, he called me two days before my trip and we talked about the work and Zewail City. Although he was very tired, he had promised to meet me in Egypt a week after my arrival to visit the lab and the campus. The next day after my arrival to Egypt, I went to the new campus. They were working day and night to finish the construction as scheduled. I took some pictures and emailed them to Dr. Zewail with a detailed explanation of all the developments at the 4D lab. Two days passed and there was no reply, I thought he was still suffering from the chest cold.

On the evening of August 2nd, 2016, someone called me and informed me that Dr. Zewail had passed away. There had been many rumors about his death, so I called his secretary immediately to be sure. She confirmed it. I just couldn't believe it .... he was gone.

I hadn't felt such sorrow for a long time, not since my father's death. The next day I went to the city to meet General Salah Azazi, who was responsible for the funeral of Dr. Zewail, and Mr. Sherif Fouad, the spokesman for Dr. Zewail.

According to his will, he wanted to be buried in Egypt, so we had to wait for his body to arrive from America. After a military funeral in the presence of the President of Egypt and some senior officials, we went to the new campus of Zewail City. We held a prayer for him with the students of Zewail City and then we went to the cemetery. Ten thousand students and friends of Zewail City were walking behind us, to say goodbye to their beloved professor.

It was a tremendously touching moment and I insisted to be with him to the very last. I helped bury him and had one last look at him. This was the final scene of the journey of one of Egypt's greatest legends, who had lit up history; and his gift to the nation, Zewail City, will continue to light up the nation's future even brighter.

Fig. 7.1

CHAPTER 7

———

ZEWAIL AND I — THOSE WERE THE DAYS

Farouk Jweideh

I would like to share what I wrote about Ahmed Zewail; the world, the friend, the artist, the inventor and the human.

> "Days that gathered strange destinies of our group in the journey of life with lots of dreams. When we met after a long journey in life we had meetings and dialogues and this is some of what went through it and around. It is the leaves of the beautiful age that brought together young people, then separate them from their dreams and homelands.
>
> Then came the end of the Chevalier while still having a great asset of dreams that life did not qualify to complete. The story of his journey is that of a dreaming scientist written by a poet who shared and witnessed some of them."

Farouk Jweideh

7.1 Ahmed Zewail: The Memories of a Beautiful Time

Ahmed Zewail left and took with him what remained of the memories of a beautiful time and the dreams of a generation which — in my opinion — was the greatest generation Egypt has seen, with its golden age culture, affiliation and love for this country. It is the generation that opened its eyes to see the revolution of July 1952, grew into maturity along with it, and spread its message to the farthest points of the world. It is the generation that dreamt of dignity, morality, prosperity and justice. Our generation lived in the early years of the 60s, in the age of flowers and free dreams.

During this time, the city of Damanhur was one of the nicest and cleanest capitals in the governorates of Egypt. It was headed by Wajih Abaza, one of the Free Officers. He turned the city into a thing of beauty. I remembered the clean streets, the cinema, the library and the railway station.

Ahmed Zewail was studying at Damanhur Secondary School and I studied at Omar Makram Secondary School. Only a short distance lay between the two schools. There was constant conflict and competition between Mr. Bhagat Mishari of Damanhur Secondary School and Mr. Sawan of Omar Makram Secondary School. Allied to each of them were their teachers, who vied with each other to provide the highest standards of discipline and the best curricula for their students. The honorable professors were highly educated — the Arabic teacher was a great scholar, the French teacher had graduated from Sorbonne University, and the Science teacher was trained abroad in several European countries.

I heard about an outstanding student at Damanhur Secondary School whose teachers were unable to outdo him in solving mathematics, chemistry and physics problems. He was named Ahmed Zewail. I got to know him and we became friends.

Damanhur Secondary School was one of the most famous schools in the republic. It competed with the best schools in Cairo — Saidia, Khediwia, and Ibrahimia — in terms of educational level, size, space and number of students, and the students' results were always impressive.

It was the habit of the people of El Behera Governorate to encourage their sons to apply to Alexandria University after their successful graduation from secondary school, because it is the closest university. However, some chose to swim against the current. I became one of those when I chose to study journalism in the Faculty of Arts at Cairo University. Their press department was the only one in all the universities of Egypt. My father — God bless him — disliked the idea because Alexandria was two steps away while studying in Cairo would entail travel, hardship and unfamiliarity. My parents mocked me, saying: "The sons of the press are people such as Mustafa Amin and other members of the establishment. How will you be able to make your way among these great press people?" I insisted I would find a way. Zewail went to Alexandria University, so we were separated at the train station.

After graduating from university, I worked at Al-Ahram Association, the great press association. Ahmed joined the Faculty of Science at Alexandria University as a Master's student. We rarely met.

In 1967, the storm of the Six-Day War broke and our generation was displaced. Zewail jumped ship and traveled to America on a scholarship. In fact, the setback was the biggest disaster our generation had experienced. It was a shock that rudely awakened us from our dreams and dealt a blow to our dignity. It caused us considerable frustration at the beginning of our adult lives. Many Egyptians tried to emigrate, but they rarely found places to study or work in Europe or America. There are those who went to the other Arab countries, where the future was certain because of the promise of oil; and there are those who insisted on staying because they could not find any other alternatives or because they clung on to what remained of their crushed dreams.

Years passed and we did not hear anything about Ahmed Zewail until he surprised us by being awarded the King Faisal Prize of Saudi Arabia in 1989, when he was 32 years old. The King Faisal Prize was a prestigious award to win, and all the more so because he was young at that time. Ten years later, in 1999, we were in for another big surprise. This time, Zewail had won the Nobel Prize in Chemistry. He was then only 43 years old.

Zewail's star started to shine — not only in Egypt, but in the whole world. He became known as the young scientist who had made the latest discovery in the world of science —— a new laser technology capable of producing light flashes just tens of femtoseconds in duration.

Fig. 7.2   Farouk Jweideh attending the ceremony of awarding Dr. Ahmed Zewail the Ordre National de la Légion d'Honneur (Chevalier), decreed by the President of France (2012). These photographs were donated by the Ambassador of France in Cairo.

Ahmed Zewail showed the Egyptians that they too are capable of remarkable scientific and human achievement. His journey began with dreams of changing Egypt, which are now a reality. His opinion was that the starting point is education. He did not know the standard of education in Egypt at that time. The last school he attended there was clean, complete with a library, a playground, a music hall and a drawing studio. There was a hot meal for all the pupils in the school every day. This was the school Zewail knew before he left.

The state institutions welcomed the successful son of Egypt with a warm welcome, the way Egyptians do — with the highest decorations and by naming streets and schools after him. He was invited to present lectures nearly everywhere. It must be understood that this young man had won the most prestigious award in science — the Nobel Prize. Zewail was very much in demand and, in a short period of time, he was as famous as the movie stars from Hollywood.

In a big ceremony, the foundation stone for Ahmed Zewail's dream — the Zewail City for Science and Technology — was laid down in Johina Square on October 6. For several years after the stone-laying ceremony, Zewail visited the site frequently in order to stand beside his dream in the desert. However, because of the old Egyptian bureaucracy, no governmental funding was allocated to build the place. Zewail was slowly losing the desire to carry on with the project. All his dreams were sealed in a vacuum despite his noble intentions, which were to improve the future of his homeland and the fate of future generations.

In these trying times, we became very close, Zewail and I. Whenever he came to Cairo, we would meet and pass the time chatting. Often, we would also meet with our colleagues from Damanhur. One of them was Mustapha al-Feki, who is now the chairman of Bibliotheca Alexandrina. Science, ideas, and poetry were the subjects of our talks, which took place at the old part of Cairo, at Al Hussein District, also known as Rehab Al Hussein. As if we had all the time in the world, we would roam around the old coffee shops at Khan El Khalily — also known as El Feshawy — and talked about anything and everything till dawn.

We were always keen on visiting Professor Mohamed Hassanien Heikal — the chairman of Al-Ahram Association — the most widely known and prestigious association, not only in Egypt but also in all of the Middle East. On the last of these visits, Professor Mohamed invited us — Zewail, Farouk Alokdah and I — for lunch at his home in. It was a day rich with dialogues, questions, and comments.

I told Zewail, "I am a man who is good with words but science shapes minds. You are the present and the future. Now is the time of science, not the time of poetry. I am part of the past, which we call our heritage. I have accepted this."

"But what about the prize money?" he asked.

"This is a non-negotiable topic." I told him to share it, but his reply to me was, "What about my share in this agreement? It is devoted to something else." He had a good sense of humor in a perfect Egyptian way.

The time he had spent in the US had not changed his behavior. He was an Egyptian to the bone and loved his old friends. His relationship with other Egyptian scientists was always warm and tender. Professor Lotfia El-Nadi, Dr. Magdy Yaqoub, Dr. Ahmed Okasha, Dr. Mohamed Ghonaim, Sayed Yassin, Farouk Elokda, and Mostafa El-Feky could and would all testify to this.

Fig. 7.3   The author, Farouk Jweideh, Dr. Ahmed Zewail and Mohamed Hassanien Heikal at the Al-Ahram Media Center.

He asked me a lot about my moments of creativity and the role of reason in creative endeavors. In turn, I asked him a lot about the issues of genes and the scientific progress mankind has achieved so far. He said to me one day, "Did you know that the brain is the only part that is difficult for scientists to understand so far? If I could turn back time, I would study brain cells because they hide the mysterious secrets of God Almighty. The functions of all the other components of the human body have now become clear to us due to science. The heart, liver, and ears have been thoroughly studied, but the brain is the secret of secrets."

I was very fond of the conversations Zewail and I had about science. His ideas were always fresh. They provided me with new perspectives, and were often deep and persuasive. However, when politics was causing us problems, resulting in crises and anxieties, I saw a different side of Zewail altogether, especially since his dreams were often politically blocked without clear reasons.

Fig. 7.4   The author, Farouk Jweideh, posing with Dr. Ahmed Zewail and Dr. Ahmed Ghnoneim (far left), the initiator of the Kidney Center, Faculty of Medicine, Mansoura University.

The revolution of January 25, 2011 saw the the old system of corruption and tyranny being overthrown. Ahmed Zewail was back at the scene again. A decision was made by the new government to resume Zewail's project. In the hustle and bustle of the January Revolution, my friend and colleague, the late Labib al-Seba'i, decided to celebrate Ahmed Zewail and all of Egypt's great sons in a beautiful celebration at Al-Ahram. Ahmed Zewail was able to rekindle his dream of initiating the project known as the Zewail City for Science and Technology. The project restarted.

My last meeting with Ahmed was one of surrender to the will of God. There was an air of optimism surrounding our meeting, because the doctors had decided to treat him with the latest stem cell technology. We talked a lot about his dreams for his new city and how it would be one of the most important features of modern Egypt. He saw a new future for Egypt. According to him, Egypt's promising youth would be afflicted with its history, but soon the rich, brilliant and talented elites will come.

In the months before Ahmed Zewail's passing, I suddenly missed his voice and started asking friends about him, telling them that he was not responding to me. During the same period, Ahmed bought a burial place and, in his will, indicated that he wanted to be buried in Egypt. On August 6, 2016, Egypt received the body of her son with much gratitude, love, and appreciation. He was accorded a state funeral with military honors led by President Abdel Fattah Sisi. A second funeral was held at the Zewail City campus, where thousands of young Egyptian students and colleagues attended to bid their final farewell to the man who was a symbol of hope and their motivation for working towards becoming a great scientist.

In his speech at the celebration of the expansion of the Suez Canal, President Abdel Fattah Al-Sisi stressed the importance of completing the project that Ahmed Zewail had started and of making his dream come to fruition. I was confident that the Egyptian army would fulfill Zewail's dream because it was the nation's dream.

When Ahmed Zewail passed away, he took away with him the age of innocence, but not the dream which he had clearly visualized and had died defending. There are many types of martyrs and Ahmed Zewail was one of them.

7.2 Zewail and the Floor of the Enchanted!

Zewail's death was among the many events that have caused Egyptians a lot of anxiety and pain. A true son of Egypt, he brought joy to his country from outside its borders. His achievements and efforts dispelled the clouds of frustration that had cast shadows over our souls. Like a light illuminating the darkness, he brought us confidence; for that was how Dr. Ahmed Zewail, the Egyptian scientist who had been awarded the Nobel Prize in Chemistry, was received by the Egyptian people. He was the cause of great celebrations on the streets of Egypt, winning the loyalty of even the simple folk. He was a beacon of hope during difficult times and a ray of divine light sent to inspire us until his journey through life came to an end.

Ahmed Zewail's winning of the Nobel Prize was a major event by any standards. This may have been due to more than one reason: In the eyes of young Egyptians, Ahmed Zewail was a tireless contributor to his country. His many successes made him a role model by showing Egyptians that any simple, young person with a dream could make something of themselves. He built a great edifice of excellence and superiority with determination, both culturally and scientifically. His very existence gave the Egyptian youth new hope, as it was indicative of the drastic social changes that Egypt experienced in those years. Previously, chasing one's dreams had been a very difficult task for ordinary people. With Ahmed Zewail as an example, many young Egyptians began to believe that no matter what social class one was born in, it was possible to achieve great things.

Hence, Ahmed Zewail's return restored the confidence of young people in the future. It did so in the face of a society that had lost its balance to the extent that everyone in it dreamt of only one thing to do — business.

The Egyptian youths of this era have been exposed to many negative influences in their formative years. All around them, they see the ideals of high art, financial success, and polite behavior that have been distorted by the realities of censorship, fraud, and sycophancy. The rise of Ahmed Zewail proves once more that it is possible for one's life's work to lead to success, even if one doesn't belong to a rich family. One doesn't need to be the son of a senior official, owning companies and hotels, or become a star in the world of football or art. In fact, an important part of Ahmed Zewail's legacy was that he gave young people an example to follow, which led them away from the traditional paths to success. Instead of dabbling in politics to achieve his ends, he devoted his life to science, after which his worth was eventually recognized by important officials. His example stood in contrast to that of the dozens of bad role models the Egyptian youths watched every day on television and read about in the newspaper.

Ahmed Zewail represents a generation obsessed with the history of modern Egypt. We call this generation the generation of reincarnation. The people of this generation were profoundly affected by the 1967 setback and came out of it with a sense of having lost their balance, even as they strode through the streets of Cairo while the enemy aircrafts flying above them obscured the sky. Ahmed was among the sons of this generation. He smuggled himself, trying to restore his balance in an unfamiliar land. This generation was named the "Generation of the Enchanted Floor" by our mutual friend Dr. Mustafa al-Faki. I do not know why he named it such. Perhaps he was thinking of how fragments from the floors of the high-rise buildings in Egypt broke off and fell to the ground during the 1967 war. Or perhaps he had found the most appropriate way of describing a generation that had experienced war, but had embraced peace, an ideal foreign to their predecessors. Torn between these conflicting influences, they seemed to float, suspended, between earth and sky.

Strangely enough, this generation, which is now in their 50s, likes to be called the generation of youth. This generation has remained this way since time reversal and even now awaits chances that, it seems, may never come.

Zewail was one of the sons of a forgotten generation. If we were to ask ourselves about Ahmed Zewail, what if he had stayed on in Egypt? What would his fate be? There is more than a possibility that he would have remained in the Faculty of Science at the University of Alexandria. Maybe he would have gotten a doctorate. Perhaps, the prevailing conditions would have led to the absence of a professor who could supervise him, or a lack of funds for travel. Perhaps, the bureaucracy would have refused to cooperate with Ahmed Zewail. Perhaps, he would have been constrained by the dearth of study options and the obsolete laboratory equipment, especially as he had chosen to study the laser, a very modern specialized field. Perhaps, Ahmed Zewail would have remained on the sidelines of life, or tried, as the sons of his generation did, to join the Left so that he could find some protection under the banner of socialism; or to follow unorthodox religious teachings, so that he could escape from this world to the next and forget his broken dreams.

It would have been possible for Zewail to enter any of these dark cellars if he had remained in Egypt, but God had chosen him for another purpose. Ahmed Zewail did not succeed in the field of ordinary experience, but in a field that is rarely studied in developing countries, because the science involved is still too far removed from the everyday lives of its citizens.

Science is still a strange way of thinking for many of Egypt's citizens, who have lived their lives steeped in backwardness, ignorance and superstition. Because science is an austere and narrow road, it does not suit those who walk myriads of meandering roads to reach their.

Science is still rarely used in the planning of our lives despite the importance of improvisation based on scientific knowledge as a problem-solving strategy. As a result, there are dozens of scientists who died without anyone hearing what they had to say and, thus benefitting from their intelligence.

Science is still strange in a society that only admires the stars of the artistic and sporting worlds, and in which songs about Egypt's intellectual decline are common.

In Egypt, we still measure our superiority by examining our checkbooks and bank balances, even if these assets of ours come from unsecured loans. In other parts of the world, superiority is measured in terms of scientific progress.

Science is still strange to us Egyptians, so we allocate lean budgets for scientific research that do not compare with the amounts we spend on shallow TV serials and dance festivals and football matches we could do not win.

Scientific research is still strange to us, so we neglect to honor the dedicated scientists who have spent their lives in research and study. State awards are given to those not worthy of them, such as senior officials in the state who are not content with their positions, but want the honors as well.

Scientific research is still difficult in Egypt due to run-down laboratories, where devices do not work and conditions in universities; schools and research facilities were not conducive for research. At a time when the world was making a new scientific breakthrough every hour, Egypt was getting left behind.

Scientific research still does not inform the way we think and deal with fundamental issues. Furthermore, we have not been able to talk about science and scientific research for the past 50 years because we are still facing the problem of illiteracy.

Following the logic of the current age and the dictates of progress, the scholars of Egypt should be in the topmost social classes, first among the elite. They should have the final word in all our affairs, from the administration of schools to the concerns of the peasants. As it stands, slogans, songs and ideas consumed our people and drove them to end their lives. Despite this, they are still heard from all our channels and platforms.

Noble Ahmed Zewail brought the plight of the youths of Egypt to our attention. This young man, who left us a long time ago and who did not have an upbringing common at that time, was puzzled. He was trying to find a way to transform our youths in order to prepare them for the new era of knowledge. Unfortunately, the youths of Egypt do not dream of participating in the future. They do not wish to play an active role in the country's progress.

Ahmed Zewail is an honorable model for the Egyptian youths. He represents a great example of how one's efforts and perseverance can pay off. Unfortunately, he was ignored but it did not hurt the man, nor diminished his ability and success. It was indicative of the general climate in Egypt at that time, which was a shame. He and others have tried, but failed to give young people real opportunities for creativity and excellence.

One must be serious and be insistent on raising the profile of science in one's country, that is the lesson Zewail wanted to teach our youths. Our youths should learn from Zewail's experience to develop themselves. They should not be made to feel that success is an impossible dream. Everyone who seeks what Zewail had achieved should leave his homeland, but to return to contribute.

We are now on the threshold of a new phase where the State is trying to raise the flag of science through action taken by the Ministry of Scientific Research. The State is also trying to address the issues of youth through action taken by the Ministry of Youth. For instance, there are new plans for the dissemination of scientific culture and the teaching of practical thinking. These developments will open up more opportunities for young scientists in Egypt.

Ahmed Zewail's success has delighted me on a personal level. As youths, we were together on the streets and in schools of the old city of Damanhour in the early 60s. We lived in our dreams and dreamt of the many ways and paths we might take. Time was innocent and so were we.

We had gathered for coffee with El-Messiri and at the Cinema City. And the sons of AbuRish and Ibrahim Aldsouki were there with us. May Allah be pleased with them.

Ahmed Zewail pleased me in other aspect as well. He symbolised the beginning of a new era in which the language of science has its place in our country — in our work, research, thinking, study and behavior. Only then, can we truly make an informed choice about which path in life to take.

I was happy, above all else, that a young man of Egypt won a Nobel Prize in science at a time when this prestigious award is being monopolized by the developed countries.

The most beautiful moments of happiness came as a coincidence. Zewail's winning of the Nobel Prize was the most beautiful coincidence that pleased the entire nation. It must be said that his achievements were not by chance. He had to pay a dear price and made great effort. I would like to pay tribute to Ahmed Zewail — a colleague and friend.

7.3 With Ahmed Zewail

One of the most beautiful things in life is to be able to make your dreams realities and to see fantasies become facts. The difference between the scientist and the poet is that the dreams of the scientist can become a reality one day and will bring us closer to an understanding of the universe. The dreams of the poet are always ephemeral. If most of the people in the world dream of things that can be touched, the poet can only experience their dreams as a light that only they can see and impossible to touch. There are many links between science and poetry, but they are difficult to identify. The world keeps the dreams of the scientist in his hands, but keeps those of the poet in the wind.

I was happy all week after having the chance to have more than one meeting with our distinguished scientist Dr. Ahmed Zewail. We met in the arena of the Al Ahli club during a beautiful celebration, held at the beginning of a large festival in the Ahli centennial, which was skillfully organised by Hassan Hamdi. We then met on the bank of the Nile with our advisor Abdel Majid Mahmoud, the Attorney General, and Dr. Mustafa al-Feki. Then, we went to Al Hussein and spent ages at the coffee shop Al-Feshawi, reminiscent of our youthful days. Cairo's charming nights passed surrounded with its old buildings and good people.

We remembered our little dreams in the early years of our youth as we strolled through the ancient city of Damanhour and imagined a new era. I walked through my beautiful homeland with this amazing man. Ahmed Zewail had not changed in any way; he still carried himself with that same simplicity and tolerance, which seemed to bridge the past and the present. He was still as curious, asking astonishing questions, despite having led humanity to the edges of knowledge. He was still enamoured by the love of this country, so much so that he would constantly pursue his dream to increase opportunities for its people as long as his heart was still beating. Despite all his achievements, he could not change the world, but he still dreamt that his countrymen will have a more prosperous life, a more justifiable time . . . and that Egypt will become a society that gives as much as it takes.

Ahmed Zewail still imagined a society that opens doors to all its members without distinguishing between old and young, between citizens and immigrants, and between those who own everything and those who have nothing. But that time has passed us by and will never return.

I have known the story of Ahmed Zewail since he was awarded the Nobel Prize and his return to Egypt with open arms. I watched his attempt to advance science and knowledge in Egypt. He met President Mubarak after his Nobel Prize and the President honored him with the highest Egyptian Medal, the Order of the Nile. The President issued clear instructions to the government at that time to listen to Ahmed Zewail and to share his dreams for Egypt. The Egyptian President Mubarak was honest and clear, and issued all these instructions to provide Ahmed Zewail with all the support he would need. With his support, Ahmed Zewail conceived a project to establish a huge scientific university that would fill the terrible dearth suffered by Egypt in the field of scientific research.

The Zewail project aimed at establishing a scientific base that would place Egypt on the doorstep of contemporary life with all its challenges. Zewail tried to gather a constellation of the world's greatest scientists — Nobel Prize winners in all areas — to show them a vision of the future of scientific research in Egypt. He did not forget to include in this constellation, a number of colleagues, teachers and students from Egypt and the Arab world.

Zewail had made long journeys between Cairo and the capitals of the world at his own expense, gathering information from here, and studying there in order to implement his dream for the people of his homeland. He fiercely resisted the interference of others in this matter, and set out with enthusiasm to battle the bureaucracy of the old Egypt. In doing so, Ahmed Zewail found himself up against a structure that was difficult to deal with on both the human and moral end. The question asked by our great Egyptian government was how the necessary funds could be acquired to finance this project. Their solution was as strange as it could possibly get. Soon, Ahmed Zewail discovered that he was required to provide funds to realize his dream for the people of his country. He answered, "I can gather scientists for you, but as for raising funds, this is not my responsibility and I am not qualified for that."

Ahmed Zewail sadly withdrew. The bureaucrats of Egypt wasted a great opportunity presented by one of her loyal sons. I did not hear this story from Ahmed Zewail but from my sources. Ahmed was surprised that I knew everything. However, I did not mention other painful details I had heard to him, including how the beautiful villa in the Garden City that had been restored by Ahmed became an excuse for senior government officials to take up a case against the man.

Qatar asked Ahmed Zewail to head a team entrusted with setting up one of the most prestigious scientific universities in the Arab world. The project's cost $3 billion. Ahmed Zewail declined the offer.

I do not know how long Egypt will be guarded by those who persecute the talented, so that those in power can remain in control of everything. It has been years since Zewail won the Nobel Prize and yet nobody asked what the project was all about. It was as if we could see swarms of crows flying in the sky, chasing every singing bird that tries to teach people how to sing in this country, while bats still scream in our skies.

In front of the Mosque of Hussein, may Allah be pleased with him, we — Zewail, Mustafa al-Fiqee, and I — sat recapturing the image of a home which contains beautiful things. I saw the tolerance of religion and its moderation, the art and its message — things that raise the souls and enrich them. We were not rich, but we had something beautiful called conviction. The lure of money did not tempt us; rather, the glamour of thoughts and culture inflamed our imagination. Our houses were simple, but love covered our walls. We had little money, but we knew something called 'pond'. Our streets were not clean, but our hearts were pure. We never read in the news about a son who killed his father, or that of a wife who disposed the remains of her husband. Is our society still pure, or have we become slaves of money now?

Ahmed Zewail said, "A few years ago there was an uproar in America saying that we are a nation in danger. It is now necessary to look at the many areas that require improvement, especially education. I say that Egypt's initial problem now is education. It is not important to establish dozens of regional universities without professors, or to accumulate thousands of students in a small room. There are only three universities in Egypt which are recognized internationally: the Universities of Cairo, Alexandria, and Ain Shams. No one will dare to question the qualifications of the graduates of these universities or to ignore their scientific value.

The second thing we must accomplish is to open the doors of opportunity to excellence and merit. It is unfair to give an opportunity to those who do not deserve one, it deprives the truly talented and deserving of a place.

The third thing is that we must look at the future and what Egypt's image will be like after 20 years. This requires us to review everything, to lay the foundation for the image that we want to cultivate. We have all the possibilities in order to become effective and influential. There will be no future without knowledge."

It was a pleasure to sit in a dreamer's world, particularly when his dreams are not for himself, but for his homeland. Even though Ahmed Zewail lived far away from his native country, he followed his homeland's development. He still carried in his heart everything that is beautiful and wonderful about Egypt. Ahmed Zewail whispered in my ear that what he was most afraid of was that Egypt will not have a place in this new universe, which is being shaped by knowledge, science and the future. To avoid this, we should overcome our problems and conflicts, and ignore the absurdities of our battles and think of one thing only — the future that awaits us in this new world.

I said to Ahmed Zewail as we shook hands, "It seems that you migrated in body, but that your heart belongs to Egypt and that you have the spirit of al-Husayn. May Allah be pleased with him."

We then roamed around the city, and it seemed to me that the call for prayers at dawn had shocked the old Cairo and shackled its silent pillars, so much so that authentically good people were drowned deeper in their deep sleep.

I had a question that needed answering. When would we be able to cut down the fruit trees and chase the beautiful birds and celebrate by opening all the doors for the bats to escape?

Ahmed Zewail traveled back to the States, I promised myself to ask him if there was a way to regain our dream again.

7.4 Before the Withdrawal of Ahmed Zewail

Sometimes, I asked myself what Ahmed Zewail has done since his return to Egypt after winning the most prestigious international award: Has he taken up a position of power and looted public money? Did he accept gifts of state land and sell it in auctions to earn billions? Ahmed Zewail became a cancer patient at the height of his success.

This story needs a great specialist, like Dr. Ahmed Okasha, to analyze what happened.

I am not talking about buildings, land or verdicts. I am talking about a simple Egyptian man who left this land, but remained true to its roots. He set off alone in this vast world, and achieved an unprecedented scientific and human achievement. He won the biggest international prize. The whole world knows that Ahmed Zewail spoke with loyalty and gratitude about the land that gave birth to him and nourished his dreams, excellence and nobility.

When Zewail won the Nobel Prize, birds flew in the skies of Egypt, and songs were sung by the Egyptians in love and appreciation. Zewail was celebrated by Egypt for a long time and the man, thankfully, was grateful for the honour. His country had gained the experiences, knowledge and achievements Zewail accumulated. He did not seek anything, not wealth nor any other gains, after he achieved success in his scientific career. He was placed in the list of the world's leading scientists in his 50s. He has enough money and more fame than he could have ever imagined. Is there anything greater than humanity's appreciation for the knowledge of those who have exceeded all the limits of knowledge?

Zewail had no other goals, but to serve the land in which he was born and grew up. This remained true even after his dreams had become a reality.

Zewail struggled to work with the Egyptian administration, which suffered from bureaucracy and ineffectiveness. Despite this, Ahmed Zewail envisioned a grand scientific project that would restore Egypt's role as a regional leader and take advantage of all its human resources and international relations.

Zewail wanted only one thing: that one day, Egypt finds its place alongside the developed countries as their partner in shaping modern civilization.

Despite all these, the state chose a piece of distinctive land in the heart of the city to be the nucleus of the project of the scientific city that Zewail had dreamed of.

Zewail began to shuttle between the lands of Egypt and USA in order to realize his dreams. Thus, began a journey of frustration, which turned the heart of the man against the power of apostles and hypocrites. Politicians chanted that he was seeking power and that he wanted to be the president of Egypt. Conspiracies began to surround the Zewail project. Despite the difficulties, Ahmed Zewail managed to leave a small memorial stone in the City, on which was written his dream of rebuilding Egypt.

Zewail withdrew from the scene quietly but, from time to time; he came to visit Egypt to meet old friends.

The January Revolution of 2011 saw the return of Ahmed Zewail's dream; thus, began a new journey to revive it. Despite the initial setbacks, Ahmed succeeded in starting a large campaign for donations. Zewail's project was given full protection by the Egyptian state. This protection extended to the land where he once planted his old dream — the city of October 6.

The Egyptian state restored the Zewail project, for which the land was allocated 15 years ago. It began the donation campaign that would raise the funds necessary to build the city. The Central Bank of Egypt issued a special law for this scientific city, which allowed the import of scientific equipment worth 300 million pounds into the city; and which allowed the entry of 300 students into Egypt as the first batch of workers to the new city. The entry of five foreign Nobel Laureates, along with that of dozens of Egyptian and foreign scientists, who had joined the research team, was also allowed. Up to that point, all the state institutions supported Ahmed Zewail's project in any way they could.

At certain points, Ahmed Zewail found himself not getting as much support as he was used to. He did not have any say in the allocation of land for his project and was facing some serious problems, but at the end, he won the respect of the people and the administration.

However, all the conflicts took a toll on Ahmed's health. Doctors told him his ailment was the result of psychological pressure and frustrations in the face of the conflicts. Ahmed Zewail's voice became softer and softer:

> "Was I mistaken to have such a dream? I have no personal interest at all, but I started this project for the future of this country. For 15 years no Egyptian institution has come forward to assist. I paid for all my trips to and fro Egypt. I do not need to have a political or administrative position; neither do I need money nor fame. My doors are always open to anyone who needs help. Now that I live in the shadow of my disease, I shall withdraw in peace, leaving this project in the hands of the Egyptian state, which has sponsored the project because they have every right to do so. When I hear the voice of my conscience berating, I tell myself that the Lord knows best."
>
> Ahmed Zewail

Ahmed Zewail prayed for recovery, but was not granted. To my mind, this can only mean one thing — the mortal realm filled with petty problems was not worthy of Ahmed Zewail; thus, he was taken away from us. He left us peacefully, and now resides in the land of the Lord.

The Egyptian state should protect the Zewail project, as promised by El-Sisi, the President of Egypt. Egypt's real crisis is that there are those who seek to raise standards in our country so that we can survive in the modern world, and yet there are also those within us who would fight for self-interests rather than for the best interests of their country.

Fig. 8.1

CHAPTER 8

GALILEO OF THE TINY — THE BRILLIANT AHMED ZEWAIL

Yehea Ismail

When Galileo Galilei (1564–1642) expanded on the work of Arab pioneer scientists in optics — such as Ibn al-Haytham — and built the first modern telescope with two lenses and a long tube, he changed astronomy, natural philosophy, our perception of the world and our position in it, and even the fundamental dogmas. It was believed that the earth is flat, and that the sun rotates around the earth with the earth and us at the center of the universe. However, by simply looking more closely at celestial objects for the first time with the Galileo telescope, it was easy to see that the earth orbits the sun and that we are not the center of the universe. Further advanced telescopes have shown we are just a tiny speckle in a vast universe of galaxies, each with hundreds of billions of stars like our Sun.

A little known fact is Galileo even perceived the principle of relativity, which Einstein later adopted and modified for objects moving at great speeds approaching the speed of light. All these changes our perceptions of the universe, our position in it. Furthermore, advances in experimental and theoretical astronomy were initiated by Galileo's invention of the telescope as a tool to look into the large universe.

Dr. Ahmed Zewail was one of the pioneers in inventing a similar device that looks into the microscopic world, which is expected to change our perceptions of science and the world as much as Galileo's telescope did, if not more. We, as humans, can only perceive things that are in comparable scale of size and speed to ourselves. For example, if we try to see something very far away, we can. Or if we want to perceive the motion of something that moves very slow, we won't be able to see such a motion.

Tools need to be developed, so that we can perceive things of different scale to us. Similarly, objects in the microscopic world — like electrons and atoms — are much smaller than the resolution of our eyes to perceive and move at extreme speeds. For example, an electron revolves around a nucleus in an atom seven million billion times every second.

No tools were available before Dr Zewail's invention to look into this microscopic world that moves at an amazing speed. Dr. Ahmed Zewail's invention of the femtosecond microscope — or the 4D microscope — enabled humans to look, for the first time, into this amazing microscopic universe with all its wonders. Dr. Zewail's discovery and invention was not just another great invention, it was an invention that opened the doors to numerous other inventions and discoveries of a whole new world.

The 4D microscope can be simply explained as a very fast camera that is fixed on a very strong microscope (electron microscope). Electron microscopes were well known for decades before Dr. Zewail's invention, but they can only take one picture at a time; hence, only have information about the microscopic world at one instant in time. Such a microscope can be called, if you will, a 3D microscope that only captures spatial information at one instant of time.

What Dr. Zewail did was that he modified the electron microscope to add the fourth dimension of time to the information extracted from an electron microscope. In simple terms, the modified 4D microscope can record movies of the microscopic world while the traditional one can only take a photo. Given the speed with which objects move in the tiny world, Dr. Zewail had to invent a camera with a very fast frame rate — in the femtosecond time scale. For example, assuming a bullet is shot, our eyes cannot see it because the movement is too fast for our eyes' frame rate. Our eyes can only take around 30 frames per second of the field of vision. Hence, if a very fast object like a bullet crosses our field of vision in less than 1/30 of a second, no frame will capture the bullet as it moves and thus it is invisible to us. However, let's say we get a camera with a very fast frame rate, we will be able to capture the bullet as it crosses the camera's field of vision in many frames and reproduce the motion of the bullet in a slow motion movie.

This technique has been used many times in science to image a drop of rain, an insect, a wave, or a bullet. However, to image the world of electrons, atoms and tiny biological structures, a super-fast frame rate was required. A femtosecond is 1 over millionth of a billionth of a second, or a second divided 10 fifteen times. A femtosecond microscope is one that can take a million billion frames for every second. This was the invention of Dr. Zewail, achieved through brilliant modifications of the electron microscope using fundamental concepts of quantum mechanics.

For his invention, Dr. Zewail' was awarded the Nobel Prize in Chemistry in 1999. He was the sole recipient. The win gave Dr. Zewail worldwide fame unlike any other Nobel prize winners. With the new 4D microscope, for the first time ever, a chemical reaction could be seen happening in real time. We can see electrons as they bond to atoms and the time taken for the bonding to complete. This initial illustration was followed by numerous illustrations for the use of this 4D microscope in other fields like biology and semiconductor physics. Many great discoveries came out of this invention, which in turn helped improve our lives and change our perceptions beyond anything else.

By understanding biological processes in detail and experimentally, new medicines or cure techniques can easily be developed. By imaging semiconductor devices and showing electrons as they move in the devices, new theories can be formulated and better devices can be developed; resulting in better computers, cell phones, communication technology, solar cells, etc. Most importantly, by understanding the workings of nature at this very fundamental level experimentally, we can develop completely new unimaginable applications that were mere science fiction in the past or even those not imagined yet.

Richard Feynman, a very famous Caltech Nobel Prize winner, has a very famous statement: "there's plenty of room at the bottom". What Feynman meant was that there is significant space for discovery and invention in the microscopic universe. Feynman was a colleague of Dr. Zewail at Caltech, and so was Carver Mead. Carver was a major pioneer in the area of scaling down the size of transistors in electronics devices. When Carver Mead started investigating the effects of scaling down a transistor, he was surprised that everything got better in electronic circuits built out of these smaller transistors. This was contrary to the common perception at the time that things would get worse as we scaled down the size of transistors. However, making transistors smaller enabled us to cram more functions in a smaller chip, resulting in less energy per computation and much faster computations.

This technology — the scaling of the sizes of transistor — is the basis of a now multi-trillion-dollar industry per year, and even bigger on industries that depend on electronics, such as the communication industry and services. Similarly, other areas have plenty of room at the bottom. Looking at organic and biological reactions at the microscopic scale, and in a detailed manner like never before can hold the key to personalized medicine, physical-based therapy, nanomedicine, new diagnostic techniques and numerous other applications to improve the quality of our lives.

The underlying economic opportunity in the biological and medical area is also enormous. The area of nanomaterial science carries a huge promise for materials that can be used in everyday objects — in energy harvesting, optical materials, and construction among many others. These are just a few examples.

The opportunities at the microscopic scale are enormous, and Dr. Zewail's 4D invention is expected to contribute significantly towards the advancement of the microscopic scale and this in turn will benefit all of humanity.

8.1 Dr. Zewail's Moral Character — How I Met Dr. Zewail and Our First Impressions

My first real acquaintance with Dr. Zewail was through Dr. Lotfia El-Nadi, who knew me and was a member on the board of the Zewail City. Dr. Zewail had asked his board of directors to nominate some of the city's future scientific leaders who have a global reputation. Dr. Lotfia chose me to lead the field of nanotechnology. My first conversation with Dr. Zewail was via the Internet in the month of Ramadan 2011, where we talked for about half an hour. We then met in the United States in Washington DC, where we spent a few hours together and identified Zewail City's main directions.

What impressed me most was Dr. Zewail's demeanor in dealing with me and the mutual respect we had for each other. Dr. Zewail was very modest despite his fame and reputation. I have agreed from the beginning to work with Dr. Zewail in the city without any remuneration, which is still the current situation, and this was due to my agreement with the ideas of Dr. Zewail and our love for science and realization of its importance.

Dr. Zewail is one of the very few who have won a solo Nobel Prize and at a relatively young age, which is very rare. He has more than 50 honorary doctorates from all over the world. He has been named as one of the top professors by some of the world's ranking sites, not only because of his Nobel Prize but also because of other factors, such as his continuing scientific activities and the number of students awarded PhD and MSc degrees (400 Student) under his supervision, in addition to the number of citations from his research articles among other things. Nobel Prize winners — like Dr. Zewail — can go around the world giving speeches and demand high fees, and do less science. However, Dr. Zewail preferred to put all his time and effort in developing science, and helping Egypt build a project as part of the scientific renaissance there with absolutely no payment at all, even at his own expense.

8.2 The Ideology of Dr. Zewail

Dr. Zewail wrote his famous book *The Age of Science* which, as its title suggested, emphasized on the idea that science is the basis for prosperity, not only in this age and time but even in ancient times. The Pharaohs of Egypt at one time dominated the world in science, economics, society, martial arts, warfare, astronomy, engineering and medicine. Similarly, the Romans and Greeks, who dominated the scientific world after the Pharaohs, had famous scientists, such as Aristotle and Pythagoras. Also, in the era of the golden age of Islam during the Abbasid caliphate, the Muslims dominated the world in science, medicine, engineering, and economy with famous scientists such as Ibn Sina, El Khwarizmi and Ibn Rushd etc. However, these civilizations collapsed with the collapse of their scientific and ethical systems.

Dr. Zewail was fully convinced that there is no other way to rejuvenate Egypt except through the advancement of science, giving attention to scientists, and realizing their crucial importance for this country. Dr. Zewail also believed that there is no way to promote science unless we observe the ethics and the culture of science. How many times have we advised each other: to respect each other, to respect intellectual property, to persevere and to work hard? The media need to spread the culture and the importance of science to this country, and invite everyone to contribute in order to create a distinct climate for scientific institutions in Egypt. As for the idea itself, a brief contemplation in some major economies reveals a lot.

Major economies based on the automotive and the electronics industries — for example, Japan, Germany and Korea — became great through exporting electronic goods and cars; look too at countries who own information technology.

They have created international companies worth billions of dollars, such as Google, Facebook and YouTube, and information wars are dramatically changing the world. Russian President Vladimir Putin, for example, has said that the one with leadership in artificial intelligence will own the world. Elon Musk, one of the most famous American scientists and entrepreneurs, said World War III will be a war of artificial intelligence and information technology.

The fields of agriculture, medicine and energy become strategic sciences. For example, renewable energy can be used to desalinate water from the depths of the earth to secure a nation's food and water security.

It is clear therefore — with little reflection — that science is the basis of a nation's prosperity, not only in this age and time but anytime. Dr. Zewail's idea of using science as the basis of prosperity is of crucial importance.

Fig. 9.1

CHAPTER 9

AHMED ZEWAIL — OUR PRIDE

Sultana N. Nahar

The Ohio State University (OSU) has been holding the International Symposium on Molecular Spectroscopy for over 60 years and is made popular by scientists all over the world. As I was heading for coffee and donut break at the symposium about nine years ago, I stumbled upon a picture of a smiling face with dark hair on a photo stand, placed on a table right outside the break room. My eyes were fixed with pleasure as below the picture read 'Ahmed Zewail Prize' in a large font. "A distinguished Muslim personality promoting science?", I thought. I wasn't aware of Zewail's contributions then. About a year later, I was reading an interesting article in Science magazine, which elaborated on the dream project of bringing a renaissance of science, technology and innovations to the regions of Middle East and Africa through a center, and with wide support from the international scientific community; name of the dreamer surfaced — Ahmed Zewail from California Institute of Technology (Caltech). My respect and curiosity grew deeper. He was the Linus Pauling Chair professor of Chemistry and professor of Physics at Caltech, and recipient of the

Nobel Prize in Chemistry in 1999. I knew immediately — Ahmed Zewail is our pride.

9.1 The Sequences of Events Happening During Chemical Reactions

Zewail opened up the field of study for chemical reactions and is known at the Father of femtochemistry. Human eyes can typically detect a maximum of 60 snaps per second, which is known as the flickers per second or FPS. At FPS higher than 60, it will look like a smear to the eyes as the brain cannot resolve them. Therefore, the sports cameraman increases the frequency of snaps (or FPS) to catch many pictures of a fast event, such as a ball hitting near the boundary line. When the pictures are run at the speed of human FPS, we can see the slow motion movie of the actual event and get the most accurate information for any judgment. Zewail found the way to see the sequences of events happening during chemical reactions. Without these, we cannot get the whole picture of the reaction, and cannot study it precisely in experiments or develop the needed theories.

In the past, scientists could study the initial reactants and products, and thus inferred intermediate states of a chemical reaction. When a chemical reaction proceeds the transitional states, it pass too fast to be detected. Zewail came up with the idea of ultrafast laser spectroscopy to study them by using the pulses or shots of a femtosecond (10^{-15}s per shot) laser to take the pictures of the transitional states, which would make a movie for the investigation of the process. These femtosecond shots are smaller than the molecular vibration or rotation rates, and hence can freeze the motions in time.

Zewail was able to create ensembles of molecules and synchronize their motion. He would use two laser pulses, one was the pump pulse to initiate the time and trigger the reaction, the other was the probe pulse to take snapshots of the evolution of the chemical reaction. He demonstrated that rotational and vibrational coherence is the key to making femtosecond movies. He showed the connection between the chemical bonds and their dynamics.

The work brought new and precise insights on the chemical reactions and opened up the field of study to a new high, not only for chemistry but also for physics and biophysics. He remained involved in research in all the mentioned areas. With his own research group, he also developed femtosecond electron diffraction for direct imaging of the evolution and structure of biological molecules or materials changing with time. He was honored with the Nobel Prize and many other prestigious recognitions for his contributions.

9.2 My First Visit to Zewail City of Science and Technology (ZCST)

I went to Egypt for the international conference on Modern Trend in Physics Research in 2008 (MTPR-08), taking my middle school son Alburuj with me.

We were engrossed by the Egyptian Museum in Cairo, the pyramids, the Nile River and Aik Sohnna by the Red Sea. By 2013, I was well connected to Egypt and started delivering atomic astrophysics courses with computational workshops under the Memorandum of Agreement (MOA) between OSU and Cairo University. Professor Lotfia El-Nadi, the coordinator of the MOA for Cairo University, suggested me to visit the Zewail City of Science and Technology, where she served as a member on the Board.

Professor Lotfia El-Nadi herself is an exemplary female scientist, who was the founder of several important organisations and conferences; namely the National Institute of Laser Enhanced Sciences (NILES), the Topical Society of Laser Sciences (TSLS), the Modern Trends in Physics Research (MTPR), and Ultra-Fast Laser Technology and Applications (UFLTA). She has also introduced a few new areas of research in physics in Cairo University. I came to know a lot more about Ahmed Zewail from her.

Zewail was an Egyptian by birth from Alexandria and went on to study at the University of Alexandria, then to the University of Pennsylvania in the USA for his PhD. Zewail felt strongly the need to enrich science and innovations in Egypt, the Middle East and Africa as a whole. Egypt had contributed wonders to human civilization long ago. Even now, it is the leading country for those from Africa and the Middle East seeking to get a degree. In 2013, Zewail wrote, "A part of the world that pioneered science and mathematics during Europe's dark ages is now lost in a dark age of illiteracy and knowledge deficiency." Zewail laid the cornerstone of what would become the dream institute in the city of 6 October. Supported by the then President Anwar Sadat, Zewail started to recruit dedicated scientists from all over the world. His idea of building the dream institute became a national project and was named the Zewail City of Science and Technology (ZCST).

Fig. 9.2   Entrance of the Zewail City of Science and Technology in 2013.

I was very much interested in visiting the Zewail City and the prospect of meeting Ahmed Zewail himself.

In 2013 when I first visited Zewail City, the institute was already in operation, although not fully, with Zewail as the Chair. I was greeted by the beautiful blue glass façade of the Administration building, and inside were modern labs with new equipment.

Fig. 9.3 The author (sixth from right) with Vice Chair Prof. Salah Obayya, Prof. El-Nadi and members of ZCST after seminar in 2016.

Most of the scientists were of Egyptian origin, who have successfully established themselves abroad. Faculty members whom I met came from the US, Canada, the UK, Italy, Germany, etc., who all have quit their jobs in their respective countries to respond to Ahmed Zewail's invitation to join the ZCST. There was even one nanotechnology specialist, who came from US and joined the ZCST with a huge grant from the US.

In the following week, I gave a seminar on my project on solar opacity and abundances. I noticed there were a few astronomers in the seminar. I donated some astronomy textbooks that were being used in US universities to the introductory courses on astronomy at the ZCST. Unfortunately, during my visit, I was not able to meet Zewail as he was traveling back and forth between Caltech and the City. His travel schedule could not be made known for security reasons.

However, I had the pleasure of seeing his grand effort, the dedication and enthusiasm going around on inside of the Zewail City. I wrote about my experience in a news article published by the American Physical Society (APS). The Zewail City of Science and Technology now has three sectors: the university, research institutes, and a technology park. The research institute sector has a number of centers: Aging and Associated Diseases, Excellence for Stem Cells Research and Regenerative Medicine, Genomics, Nanotechnology, Nanoelectronics and Devices, Materials Science, Imaging and Microscopy, X-Ray Determination of the Structure of Matter, Fundamental Physics, Photonics and Smart Materials, Economics and Development, Learning Technologies, etc.

9.3 Missed Meeting Zewail at ZCST Again

In 2015, I visited the ZCST again, this time to give a seminar on X-ray spectroscopy for effective cancer treatment. I missed meeting Zewail again, but instead met the person-in-charge, Vice Chair Prof. Salah Obayya, who oversees the daily business at ZCST. He was appointed by Zewail. Prof Salah Obayya is a very dedicated physicist, who left his position in England to join the ZCST. He had made considerable contributions in photo-electronics and was the author of two elaborate textbooks. Zewail City had already incorporated its university component then.

However, in 2013, a legal tussle over the ZCST land almost derailed Ahmed Zewail's dream. The Supreme Administrative Court ruled in favor of Nile University, affirming the institution's ownership rights to the lands and buildings. The name, Zewail City of Science and Technology, was no longer on the beautiful blue façade of the Administration building. It was a big blow after years of extensive and thoughtful efforts in working towards the renaissance of science and technology in Egypt. The Egyptian Government came forward by giving ZCST a new and larger piece of land in 6 October City and promised to speed up construction of the institute under the army's supervision. During that period, ZCST had to run its operation at the buildings at the back of its previous premise until it was able to move to its new buildings after construction was completed. Zewail was already familiar with administration and global politics on education. He had served as Science and Technology advisor to the US President Barrack Obama and was his appointed US Science Envoy to the Middle East.

9.4 Zewail Honored by the Symbolic Plaque of Aligarh Muslim University of India March 2016

OSU was a sponsor and partner of the IV International Conference on Nanoscience and Nanotechnology (Aligarh Nano-IV) in 2014, organized by Aligarh Muslim University (AMU) in India. I was the convener from OSU. The chief convener, Professor Alim Naqvi, the founder of the Nanotechnology Center at AMU, asked whether we could bring someone prominent in the field as a highlight of the conference. I remembered talking to an Egyptian academic, who suggested Prof. Ahmed Zewail. Based on that suggestion, I invited Prof. Zewail to Aligarh Nano-IV conference. Zewail could not attend, but he sent a welcoming message to the conference participants that I read it at the inauguration ceremony. His message was the highlight of the conference. I invited him again in 2016 for Aligarh Nano-V, organized jointly with the STEM Education and Research of OSU, and held at AMU in India. He was not able to attend, but sent the following message:

> ". . . advances in nanoscience and nanotechnolgy are increasingly impacting our lives and society. I am pleased that Aligarh Muslim University is now playing a significant role in India with its inventions in nanomaterials and nanoscience. I am also pleased that Ohio State University is involved in the STEM program for teaching and research in Indian universities . . ."

The press found it interesting and highlighted it in the media.

A large plaque with Stretchy Hall as the background, a symbol of AMU, was given to him in honor after I read his message in the inauguration session. I felt proud to accept it on behalf of Zewail.

Fig. 9.4   Left: The message of Prof. Ahmed Zewail was highlighted at the Aligarh Nano-IV international conference on nanotechnology in India in 2014. Right: He was honored by the symbolic plaque of Aligarh Muslim University of India at the aforesaid conference, which was held jointly with the International Conference on STEM Education and Research of OSU-AMU in March 2016. Sultana Nahar, see here, accepting the plaque on behalf of Zewail.

In early August of 2016, Prof. Naqvi and a faculty friend at Basra University sent me a sad message separately to inform me on the passing of Prof. Ahmed Zewail. A few months after his passing, I visited ZCST and felt a huge vacuum left after his passing and made worse by the downfall of the Egyptian economy at that time. However, when I gave a seminar on the spectroscopy of exoplanets to enthusiastic group of scientists and students, at that moment, I felt his spirit was still there.

9.5 Visiting Zewail's Final Resting Place

Laboratories were continuing with their experiments and students were engaged in discussions and activities the way Zewail would like them to be. Vice Chair Salah Obayya was steering ZCST in the right direction. They had found a leader who could maintain ZCST's economic flow by bringing in grants for the research, education and innovation. On the way back, Professor El-Nadi and I stopped by the tomb of Prof. Ahmed Zewail. He chose be to buried in Egypt and built a tomb for himself

Fig. 9.5   Ahmed Zewail's tomb.

not far from ZCST. It is roofless, but even with walls surrounding the structure, it can easily be seen through the iron fence. There is an open squared pavement space outside the wall where people can offer their prayer. Both Prof. El-Nadi and I offered our prayers before leaving. This place will remain a landmark for others to come to pay respect and offer prayers for his soul. Zewail was courageous in soldiering on towards achieving his goals. Even after he was diagnosed with cancer, he continued his work at Caltech and at the same time as Chair of the ZCST. To honor his effort, there is a plan to build a university near his birth place. His legacy will continue to inspire and contribute towards the benefit of humanity.

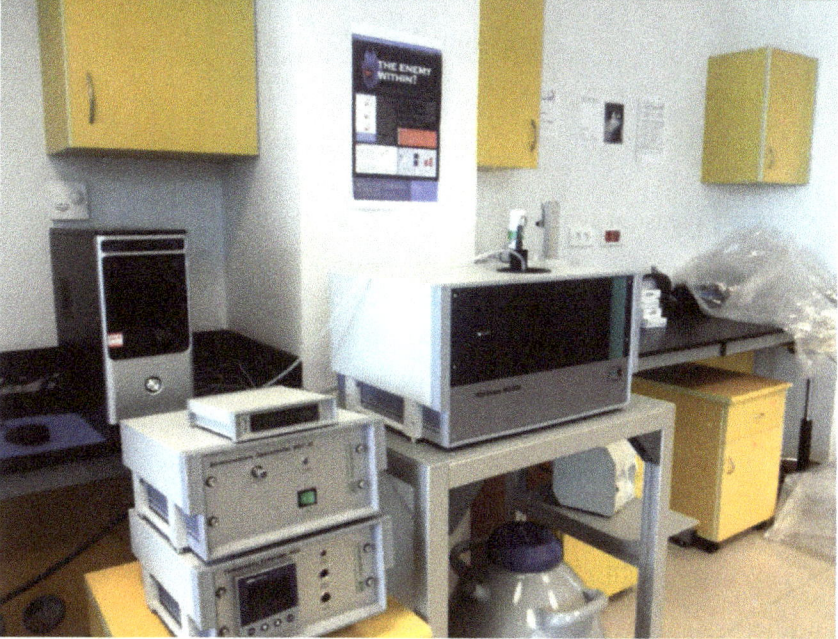

Fig. 10.1

133

CHAPTER 10

AHMED ZEWAIL — A PIONEER VOYAGER IN THE MOLECULAR SPACETIME!

Sameh Ali Saad

It was a pleasant winter day with the usual sunshine and cool breeze during February in the beautiful Southern California. My heart was beating quickly as I approached the Athenaeum, an elegant club that first opened its doors in 1930 on the campus of the California Institute of Technology, Caltech, where I planned to meet Ahmed Zewail. It was not my first meeting with him, but it was my first exclusive meeting with him.

10.1 Presenting My Fascination about Aging Research to Dr. Zewail

I was driving up to Pasadena from San Diego and during the 2-hour scenic journey, my mind was storming with ideas, fears and hopes.

It all started a few days prior to this exciting meeting when I sent a short email to Dr. Zewail, seeking an opportunity to serve my country through his recently inaugurated project, Zewail City of Science and Technology. I came to know about the project during my short annual vacation in Egypt on December 2011 from Dr. Shaaban Khalil, an outstanding Egyptian physicist, who was the first appointee to the Zewail City project. The project sounded like a wonderful path for scientific revolution in Egypt following the January 25th revolution. Immediately after returning to San Diego, I wrote an email to Dr. Zewail to explore possible ways I could help.

I am not sure what caught Dr. Zewail's attention in my email, but to my great surprise, he responded the next morning. That morning, it was the usual routine apart from the louder-than-usual wake-up call from my daughter Nour before I left for work. Her "smiley voice" called to me, "*Ezzayak Ya Sameh?*", the Arabic words for "How are you, Sameh?", as if we were old friends.

> "Good morning Dr. Zewail, this is such an honor!" I responded.
>
> "I knew you were awake, Maggie [his admin assistant] said it was too early to call, but I was sure you were awake!"
>
> "Not voluntarily!", I exclaimed with a short laugh.
>
> "I got your short message and really liked your intelligent and succinct email. Please send me your C.V. and I'll get back to you after having a look."
>
> "Right away, Dr. Zewail — I'll send it immediately upon my arrival to my office."

10.2 Zewail Introduced Laser Aided Spectroscopy to Resolve Complex Chemical Systems of Biology

At that time, I was working as an assistant professor in the Department of Anesthesiology at the University of California in San Diego. I am a formally trained physical chemist, but my interest is in free radical biology and medicine, which started in 2003 while I was exploring the extraordinary antioxidant properties of fullerenes *in vitro* and *in vivo*. The elegance and complexity of the mitochondrion fascinated me so much that I made a career shift to study the critical roles of mitochondria in aging and diseases. During my time at the Department of Medicine and Anesthesiology, I contributed to several projects in the field of mitochondria function, and free radical biology and medicine with particular emphasis on aging and associated diseases. I started my career roaming in the realm of physical chemistry, but was later captivated by the mystical elegance and chaotic complexity of biology. Compared with decisive mathematical proofs, stringent physical formulae and the predictable molecules of chemistry, biology — like life itself — has often seemed wild, unpredictable and often confusing. These challenging features which have characterized biology appeared to have become attractors for many physicists and mathematicians, including Ahmed Zewail.

I heard about Ahmed Zewail for the first time as early as 1991 when I starting my graduate studies at Tohoku University, Japan in the laboratory of the late Dr. Tohru Azumi, a prominent photochemist, who knew about the seminal work by Dr. Zewail. The neighbouring laboratory, which was led by Dr. Yuichi Fujimura, was exceptionally friendly towards me for some unknown reason. I was struggling with the Japanese language and they were struggling with English, but I managed to infer the reason behind their friendly attitude. They were collaborating with Dr. Zewail on simulations of ultrafast dynamics of chemical reactions and everyone saw some resemblance between me and him. I didn't see the resemblance. Nonetheless, I was extremely proud when Dr. Fujimura said Zewail would certainly be awarded the Nobel Prize for his pioneering work in the field of femtochemistry. Since that time, I started following his work closely and never imagined I would have the opportunity to work with him directly.

Zewail was not an ordinary scientist; his achievements and brilliance cannot simply be recapitulated by being the sole winner of the 1999

Nobel Prize in Chemistry for his "studies of the transition states of chemical reactions using femtosecond spectroscopy". Zewail's first scientific paper was published in 1970, his early work elegantly demonstrated the power and insight of laser-aided spectroscopic studies on ultrashort timescales. Later, Ahmed believed that these techniques would help resolve the more complex chemical systems of biology and materials science. His development of 4D electron microscopy, which visualizes chemical reactions in space and time, provided an extremely innovative system for the sophisticated monitoring of molecular systems in real time. Overall, Zewail and his group and collaborators produced over 600 outstanding publications in world-class journals and periodicals.

For these outstanding achievements, Dr. Zewail, justifiably, received worldwide recognition and honorary awards. He was awarded honorary Doctor of Science degrees from over 50 international universities including Yale, Cambridge, Boston, Tohoku, Oxford, Peking, Lund, Rome, etc., to mention a few. In addition to being the Linus Pauling Professor of Chemistry and Professor of Physics at Caltech; the Director at the Physical Biology Center for Ultrafast Science and Technology (Caltech), he held professorial positions at more than 25 reputable universities around the world.

In addition to his exceptional research profile, Ahmed Zewail was a very passionate communicator of science, and a vivid fighter with a strong belief that scientific research and discoveries are the only path to the future in the Middle East. These efforts were reflected in his selection by President Obama to be a member in the Council of Advisors on Science and Technology (2009–2013), and by the President of Egypt to be in the Advisory Council of Distinguished Scholars and Experts (2014). These efforts were also reflection in his selection to be in the UN Secretary General Ban Ki-moon's Scientific Advisory Board (2013). Zewail was the United States' first Science Envoy to the Middle East (2009–2011).

After sending my CV, I was not expecting a prompt response; to my utmost surprise, I received another call from him on the same day. Apparently, he liked my CV and wanted to see me in Pasadena.

10.3 The Egyptian Zewail

Upon my arrival, I saw a car with a personalized license plate reading *Musr*, the Arabic word for Egypt. I immediately predicted Dr. Zewail was already there. I was deeply moved and inspired by his love for Egypt, despite living most of his life away from his homeland. This feeling was confirmed when he took me for a tour of the campus, which included his office. In his Caltech office, an Egyptian guest would be able to relate with everything: the Umm Kolthoum's songs in the background, the Egyptian artifacts, the Arabic books and the framed Arabic calligraphies. In his Nobel Lecture, Zewail stressed his fascination with our Egyptian ancestors who contributed to the development of the "Science of Time".

In an article published in the Independent in 2006, he called on the Arab world not to be distracted by "the ideologies of the past and conspiracy theories of the future", but to forge a new "*jihad* for modernity and enlightenment". To help achieve this goal, Zewail

Fig. 10.2   Dr. Sameh Ali introducing research plan in CAAD during an early visit by Zewail City board members, including Dr. Zewail, Dr. Magdi Yacoub, Dr. Mohamed Ghoneim and Dr. Lotfia El-Nadi.

campaigned to raise money for a new science-based university and research campus in the 6th of October City near Cairo. Despite his academic honors, awards, and public recognitions, Dr. Zewail once told me that Zewail City of Science and Technology was his greatest achievement and his biggest dream came true.

10.4 My Enthusiastic Scientific Tale

I was too excited to eat my Caesar salad, which I ordered at the Athenaeum. Zewail was listening to my research narration and future plans for three hours continuously, the only interruptions being inquisitive questions. I will never forget his admiring nods and affirmations while I was explaining to him how I discovered the mechanism of action of carboxyfullerenes in biological systems. After I finished my long and enthusiastic scientific tale, he started telling me about his project; Egypt National Project for Scientific Renaissance. He was dreaming of the perfect academic city, Plato's republic of science, where we create a new culture of integrity, collaboration, and academic excellence. By the end of our meeting, he asked when I would be able to return to Egypt for good to establish a research center, but he didn't give me the opportunity to give my answer. He continued by saying, "We need to start yesterday. How about next July?"

And July 2012, it was!

10.5 The Lasting City of Zewail!

I am really lucky to have met Dr. Zewail and to have been given the opportunity to contribute to such a noble cause, regardless of the current hurdles and setbacks.

Zewail will continue to inspire me and many others because he exemplifies success in every aspect of his life, not only academic achievements.

The Greek philosopher Diogenes used to carry a lamp during the day, claiming to be looking for an honest man. I believe we don't need such a lamp because Zewail has been honestly working to improve his country and humanity. Like the stars, Zewail was bright, energetic, endearing and influential. Carl Sagan once said, "Far better it seems to me, in our vulnerability, is to look death in the eye and to be grateful every day for the brief but magnificent opportunity that life provides."

Humanity will always be grateful for the few prominent figures in history and Ahmed Zewail stands tall amongst them.

Fig. 10.3   Staff, members and students in front of Zewail City for Science and Technology — Egypt's National Project (This was the campus during the years 2011–2015).

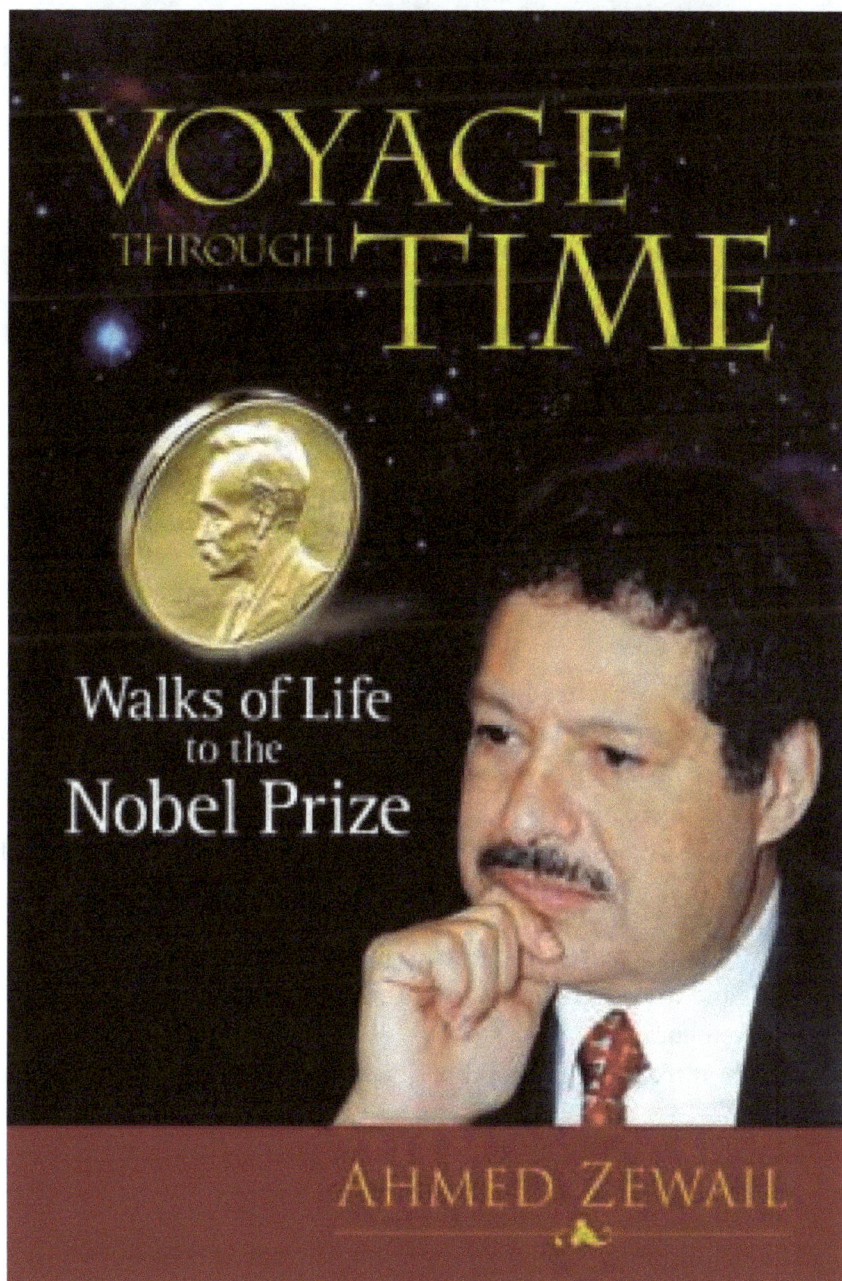

Fig. 11.1

CHAPTER 11

THE SCIENTIFIC EFFORTS OF DR. ZEWAIL

Lotfia El-Nadi

In this chapter, we introduce readers to the major scientific efforts carried out by Professor Zewail. It will be divided into three areas in which his personal efforts were significant. These areas are his scientific publications, his patents and his books, and his awards and honors. Readers will learn more about Dr. Zewail and, hopefully, by the end of the chapter they will be able to answer the following questions:

- How did his scientific interests change over the years?
- What and when were the major turning points that led to his greatest achievement?
- Which of his students stood beside him at the highest point of his career?
- How did he manage to present his thoughts?
- Which years of his life were the most productive?

His publications will be arranged by the year in which they were published. Those shaded in grey indicate articles where he was the only author. Those shaded in black indicate articles published in prestigious scientific journals (with high impact factors), such as *Science* and *Nature*. Those shaded in green indicate articles that led to major awards, like the Nobel Prize and the prestigious King Faisal Award.

11.1 List of Dr. Zewail's Publications

Excerpts from Dr. Zeweail's autobiography will head the list of publications at different stages of his life.

A.1 *Excerpt from Dr. Zewail's autobiography*

"On the research side, I finished the requirements for a Masters in Science in about eight months. The tool was spectroscopy, and I was excited about developing an understanding of how and why the spectra of certain molecules change with solvents. This is an old subject, but to me it involved a new level of understanding that was quite modern in our department. My research advisors were three: the head of the inorganic section, Professor Tahany Salem and Professors Rafaat Issa and Samir El Ezaby, with whom I worked most closely; they suggested the research problem to me, and this research resulted in several publications. I was ready to think about my Ph.D. research (called 'research point') after one year of being a Moeid (demonstrator). Professors El Ezaby (a graduate of Utah) and Yehia El Tantawy (a graduate of Penn) encouraged me to go abroad to complete my Ph.D. work. All the odds were against my going to America. First, I did not have the connections abroad. Second, the 1967 war had just ended and American stocks in Egypt were at their lowest value, so study missions were only sent to the USSR or Eastern European countries. I had to obtain a scholarship directly from an American University. After corresponding with a dozen universities, the University of Pennsylvania and a few others offered me scholarships, providing the tuition and paying a monthly stipend (some $300)."

A.2 Dr. Zewail's research works in Alexandria University for his MSc degree

1. Spectrophotometric Studies of Some Dihydroxyanthraquinones in Aqueous Solutions.
   R. M. Issa, M. S. El-Ezaby, and A. H. Zewail
   Z. Phys. Chem. **244**, 155 (1970)

2. Spectral Studies of Some Hydroxy-Derivatives of Anthraquinones.
   M. S. El-Ezaby, T. M. Salem, A. H. Zewail, and R. M. Issa
   J. Chem. Soc. B, 1293 (1970)

3. Spectrophotometric Studies on Purpurine and Quinalizarin in Aqueous Solutions.
   R. M. Issa and A. H. Zewail
   U.A.R. J. Chem. **14**, 461 (1971)

B.1 *Excerpt from Dr. Zewail's autobiography*

"My presence — as an Egyptian at Penn — was starting to be felt by the professors and students as my scores were high, and I also began a successful course of research. I owe much to my research advisor, Professor Robin Hochstrasser, who was, and still is, a committed scientist and educator. The diverse research problems I worked on, and the collaborations with many able scientists, were both enjoyable and profitable. My publication list was increasing in length, but just as importantly, I was learning new things literally every day — in chemistry, in physics and in other fields. The atmosphere at the Laboratory for Research on the Structure of Matter (LRSM) was most stimulating and I was enthusiastic about researching in areas that spanned the disciplines of physics and chemistry (sometimes too enthusiastic!). My courses were enjoyable too; I still recall the series 501, 502, 503 and the physics courses I took with the Nobel Laureate, Bob Schrieffer. I was working almost 'day and night,' and doing several projects at the same time: The Stark effect of simple molecules; the Zeeman effect of solids like NO_2 — and benzene; the optical detection of magnetic resonance (ODMR); double resonance techniques, etc. Now, thinking about it, I cannot imagine doing all of this again, but of course then I was 'young and innocent'. The research for my Ph.D. and the requirements for a degree were essentially completed by 1973, when another war erupted in the Middle East. Returning was important to me, but I also knew that Egypt would not be able to provide the scientific atmosphere I had enjoyed in the U.S. A few more years in America would give me and my family two opportunities: First, I could think about another area of research in a different place (while learning to be professorial!). Second, my salary would be higher than that of a graduate student, and we could then buy a big American car that would be an impressive acquisition for the new Professor at Alexandria University! I applied for five positions, three in the U.S., one in Germany and one in Holland, and all of them with world-renowned professors. I received five offers and decided on Berkeley."

B.2 *Dr. Zewail's research works in University of Pennsylvania for his PhD degree*

4. Optical Spectroscopic Determination of the Zero-Field Splitting in Vibronic Levels of the Triplet State of Nitrite.
   R. M. Hochstrasser and A. H. Zewail
   J. Chem. Phys. **54**, 2979 (1971)

5. Characterization of Triplet States of Axially Symmetric Benzenes Using the Zeeman Effect.
   R. M. Hochstrasser, J. E. Wessel, J. D. Whiteman, and A. H. Zewail
   Chem. Phys. Lett. **10**, 452 (1971)

6. Stark and Zeeman Effects on the Singlet $n\pi^*$ State of s-Triazine.
   R. M. Hochstrasser and A. H. Zewail
   Chem. Phys. Lett. **11**, 157 (1971)

7. Zeeman Effect Studies of the Triplet States of Benzene.
   R. M. Hochstrasser, J. E. Wessel, and A. H. Zewail
   J. Chem. Phys. **55**, 3596 (1971)

8. Studies of the 3455-Å Triplet State of *s*-Triazine.
   R. M. Hochstrasser and A. H. Zewail
   J. Chem. Phys. **55**, 5291 (1971); *ibid.* **57**, 1018 (1972)

9. Infrared Spectrophotometric Study of Some Hydroxyanthra-
   quinones and Their Sodium Salts.
   R. M. Issa, Y. Z. Ahmed, and A. H. Zewail
   Egypt. J. Chem. **15**, 1 (1972)

10. Mixed Magnetic and Electric Dipole Transition in *s*-Triazine.
    R. M. Hochstrasser, T. S. Lin, and A. H. Zewail
    J. Chem. Phys. **56**, 637 (1972)

11. Optical, Magnetic Resonance, and ENDOR Studies of the $n\pi^*$
    Triplet State of Benzophenone in Mixed Crystals.
    R. M. Hochstrasser, G. W. Scott, and A. H. Zewail
    J. Chem. Phys. **58**, 393 (1973)

12. Magnetic Sensitivity of the 3330.8 Å State of *s*-Triazine.
    R. M. Hochstrasser and A. H. Zewail
    Chem. Phys. Lett. **21**, 15 (1973)

13. Optical and Magnetic Resonance Spectra of Linear Chain Excitons.
    R. M. Hochstrasser and A. H. Zewail
    Chem. Phys. **4**, 142 (1974)

14. Experimental Studies of Triplet Exciton Bands of Molecular
    Crystals.
    R. M. Hochstrasser, T.-Y. Li, H.-N. Sung, J. E. Wessel, and A. H.
    Zewail
    Pure Appl. Chem. **37**, 85 (1974)

C.1 *Excerpt from Dr. Zewail's autobiography*

"Early in 1974 we went to Berkeley, excited by the new opportunities.
Culturally, moving from Philadelphia to Berkeley was almost as much

of a shock as the transition from Alexandria to Philadelphia — Berkeley was a new world! I saw Telegraph Avenue for the first time, and this was sufficient to indicate the difference between Berkeley and Philadelphia. I also met many graduate students whose language and behavior I had never seen before, neither in Alexandria, nor in Philadelphia. I interacted well with essentially everybody, and in some cases I guided some graduate students. But I also learned from members of the group. The obstacles did not seem as insurmountable as they had seemed when I came to the University of Pennsylvania because culturally and scientifically I was better equipped. Berkeley was a great place for science — the BIG science. In the laboratory, my aim was to utilize the expertise I had gained from my Ph.D. work on the spectroscopy of pairs of molecules, called dimers, and to measure their coherence with the new tools available at Berkeley. Professor Charles Harris was traveling to Holland for an extensive stay, but when he returned to Berkeley we enjoyed discussing science at late hours! His ideas were broad and numerous, and in some cases went beyond the scientific language I was familiar with. Nevertheless, my general direction was established. I immediately saw the importance of the concept of coherence. I decided to tackle the problem, and, in a rather short time, acquired a rigorous theoretical foundation which was new to me. I believe that this transition proved vital in the subsequent years of my research. I wrote two papers with Charles, one theoretical and the other experimental. They were published in *Physical Review*. These papers were followed by other work, and I extended the concept of coherence to multidimensional systems, publishing my first independently authored paper while at Berkeley. In collaboration with other graduate students, I also published papers on energy transfer in solids. I enjoyed my interactions with the students and professors, and at Berkeley's popular and well attended physical chemistry seminars. Charles decided to offer me the IBM Fellowship that was only given to a few in the department. He strongly felt that I should get a job at one of the top universities in America, or at least have the experience of going to the interviews; I am grateful for his belief in me. I only applied to a few places and thought I had no chance at these top universities. During the process, I contacted Egypt, and I also considered the American University in Beirut (AUB). Although I visited some places, nothing was finalized, and I was preparing myself for the return to Egypt. Meanwhile, I was busy and excited about the new research I was doing. Charles decided to build a picosecond laser, and two of us in the group were involved in this hard and 'non-profitable' direction of research (!); I learned a great deal about the principles of lasers and their physics."

C.2 *Dr. Zewail's research works in University of California, Berkley*

15. Coherence in Electronically Excited Dimers: The Observation of Coherent Dimers and Its Relationship to Coherent Excitons.
    A. H. Zewail and C. B. Harris
    Chem. Phys. Lett. **28**, 8 (1974)

16. Triplet Exciton Band Structure of Crystalline Phenazine.
    A. H. Zewail
    Chem. Phys. Lett. **29**, 630 (1974)

17. Coherence in Electronically Excited Dimers. II. Theory and Its Relationship to Exciton Dynamics.
    A. H. Zewail and C. B. Harris
    Phys. Rev. B **11**, 935 (1975)

18. Coherence in Electronically Excited Dimers. III. The Observation of Coherence in Dimers Using Optically Detected Electron Spin Resonance in Zero Field and Its Relationship to Coherence in One-Dimensional Excitons.
    A. H. Zewail and C. B. Harris
    Phys. Rev. B **11**, 952 (1975)

19. Coherence in the Excited States of Multidimensional Systems: Dimer and Exciton Dynamics in Crystalline Phenazine.
    A. H. Zewail
    Chem. Phys. Lett. **33**, 46 (1975)

20. Zero Field Optically Detected Magnetic Resonance of Multiple Localized States in 1,4-Dibromonaphthalene Crystals.
A. M. Nishimura, A. H. Zewail, and C. B. Harris
J. Chem. Phys. **63**, 1919 (1975)

21. Energy Transfer in One-Dimensional Molecular Crystals: Direct and Indirect Energy Exchange in the Non-Boltzmann Regime.
M. T. Lewellyn, A. H. Zewail, and C. B. Harris
J. Chem. Phys. **63**, 3687 (1975)

22. Optical and Magnetic Resonance Properties of the Triplet State of Biphenyl-h10 in Biphenyl-d10 and 4.2 K Neutron Diffraction of Biphenyl-d10.
R. M. Hochstrasser, G. W. Scott, A. H. Zewail, and H. Fuess
Chem. Phys. **11**, 273 (1975)

23. Coherent Energy Migration in Solids: Determination of the Average Coherence Length in One-Dimensional Systems Using Tunable Dye Lasers.
R. M. Shelby, A. H. Zewail, and C. B. Harris
J. Chem. Phys. **64**, 3192 (1976)

24. Incoherent Resonance Decay and Coherent Optical Ringing from Coherently Prepared Electronic States: A New Technique for Probing Phase Memory and Radiationless Relaxation in Gases and Solids.
A. H. Zewail, T. E. Orlowski, and D. R. Dawson
Chem. Phys. Lett. **44**, 379 (1976)

25. Optical Ringing in Solids: The Observation of Optical Coherence and Resonance Decay from Selectively Prepared Molecular Packets in Molecular Crystals.
A. H. Zewail and T. E. Orlowski
Chem. Phys. Lett. **45**, 399 (1977)

26. Laser Spectroscopy of Electronic Resonances: An Electro-Optic Technique for Probing Coherent and Incoherent Decay Processes.
A. H. Zewail

Proc. SPIE, Vol. 82: Unconventional Spectroscopy, eds. J.M. Weinberg and T. Hirschfeld, SPIE, Palos Verdes Estates, 1976, p. 43

27. New Laser Techniques for Time-Resolved Spectroscopy.
A. H. Zewail
Opt. Eng. **16**, 206 (1977)

D.1 *Excerpt from Dr. Zewail's autobiography*

"During this period, many of the top universities announced new positions, and Charles asked me to apply. I decided to send applications to nearly a dozen places and, at the end, after interviews and enjoyable visits, I was offered an Assistant Professorship at many, including Harvard, Caltech, Chicago, Rice, and Northwestern. My interview at Caltech had gone well, despite the experience of an exhausting two days, spent visiting a different faculty member in chemistry and chemical engineering every half hour. The visit was exciting, surprising and memorable. The talks went well and I even received some undeserved praise for style. At one point, I was speaking about what is known as the FVH picture of coherence, where F stands for Feynman, the famous Caltech physicist and Nobel Laureate. I went to the board to write his name and all of a sudden I was stuck on the spelling. Halfway through, I turned to the audience and said, 'Do you know how to spell Feynman?' A big laugh erupted, and the audience thought I was joking — I wasn't! After receiving several offers, the time had come to make up my mind, but I had not yet heard from Caltech. I called the Head of the Search Committee, now a colleague of mine, and he was lukewarm, encouraging me to accept other offers. However, shortly after this, I was contacted by Caltech with a very attractive offer, asking me to visit with my family. We received the red carpet treatment, and that visit did cost Caltech! I never regretted my decision to accept the Caltech offer.

My science family came from all over the world, and its members were of varied backgrounds, cultures, and abilities. The diversity in this 'small world' I worked in daily provided the most stimulating environment, with many challenges. I was very optimistic. Over the years, my research group has had close to 150 graduate students, postdoctoral fellows, and visiting associates. Many of them are now in leading academic, industrial and governmental positions. Working with such minds in a village of science has been the most rewarding experience — Caltech was the right place for me."

D.2 *Dr. Zewail's early research works in Caltech*

28. Radiationless Relaxation in "Large" Molecules: Experimental Evidence for Preparation of True Molecular Eigenstates and Born-Oppenheimer States by a Coherent Light Source.
A. H. Zewail, T. E. Orlowski, and K. E. Jones
Proc. Natl. Acad. Sci. U.S.A. **74**, 1310 (1977)

29. Laser Excitation of Stationary States in Molecules: The Coupling of Real Eigenstates to Phonons and Photons.
A. H. Zewail, K. E. Jones, and T. E. Orlowski
Spectrosc. Lett. **10**, 115 (1977)

30. High-Resolution Time Resolved Spectroscopy of Collisionless Molecular Beams: Optical $T1$ and $T2$.
A. H. Zewail, T. E. Orlowski, R. R. Shah, and K. E. Jones
Chem. Phys. Lett. **49**, 520 (1977)

31. Spontaneously Detected Photon Echoes in Excited Molecular Ensembles: A Probe Pulse Laser Technique for the Detection of Optical Coherence of Inhomogeneously Broadened Electronic Transitions.
A. H. Zewail, T. E. Orlowski, K. E. Jones, and D. E. Godar
Chem. Phys. Lett. **48**, 256 (1977)

32. Photon Trapping and Energy Transfer in Multiple-Dye Plastic Matrices: An Efficient Solar-Energy Concentrator.
B. A. Swartz, T. Cole, and A. H. Zewail
Opt. Lett. **1**, 73 (1977)

33. Optical Dephasing of Small and Large Molecules: Coherent Oscillations of Emitting Molecules.
T. E. Orlowski, K. E. Jones, and A. H. Zewail
Chem. Phys. Lett. **50**, 45 (1977)

34. Anderson's Localization of Molecular Excitons in Substitutionally Disordered Systems.
D. D. Smith, R. D. Mead, and A. H. Zewail
Chem. Phys. Lett. **50**, 358 (1977)

35. Emission Detection of Optical Coherence: Optical Dephasing and Radiationless Relaxations in Molecular Crystals.
A. H. Zewail
Proc. VIIIth Molecular Crystal Symposium, Santa Barbara, California, 1977, p. 430

36. Coherent Optical Spectroscopy of Molecules Undergoing Resonance Scattering and Radiationless Transitions: The Right-Angle Photon Echo.
A. H. Zewail
Springer Series in Optical Sciences, Vol. 7: Laser Spectroscopy III eds. J. L. Hall and J. L. Carlsten, Springer-Verlag, Berlin-Heidelberg, 1977, p. 268

37. Coherent Optical Spectroscopy of Molecules and Molecular Beams.
A. H. Zewail, D. E. Godar, K. E. Jones, T. E. Orlowski, R. R. Shah, and A. Nichols
Proc. SPIE, Vol. 113: Advances in Laser Spectroscopy I, ed. A. H. Zewail, SPIE, Bellingham, 1977, p. 42

38. High-Resolution Time Resolved Spectroscopy of Collisionless Molecular Beams. II. Energy Randomization and Optical Phase Relaxation of Molecules in Crossed Laser and Molecular Beams.
K. E. Jones, A. Nichols, and A. H. Zewail
J. Chem. Phys. **69**, 3350 (1978)

39. Measurements of Molecular Dephasing and Radiationless Decay by Laser-Acoustic Diffraction Spectroscopy.
T. E. Orlowski, K. E. Jones, and A. H. Zewail
Chem. Phys. Lett. **54**, 197 (1978)

40. Coherent Transients by Laser-Acoustical Diffraction Spectroscopy.
A. H. Zewail
J. Opt. Soc. Am. **68**, 696 (1978)

41. A Novel Laser Technique for Observing Optical Coherence in Gases, Molecular Beams, and Solids.
T. E. Orlowski and A. H. Zewail
Proc. Electron-Optics and Laser Conf., Industrial and Scientific Conference Management, 1977, p. 602

42. Coherent Processes in Molecular Crystals.
D. M. Burland and A. H. Zewail
Advances in Chemical Physics, Vol 40, eds. I. Prigogine and S. A. Rice, John Wiley & Sons, New York, 1979, p. 369

43. Molecular Mechanisms for Dephasing: Toward a Unified Treatment of Gases, Solids, and Liquids.
K. E. Jones, A. H. Zewail and D. J. Diestler
Springer Series in Chemical Physics, Vol. 3: Advances in Laser Chemistry, ed. A. H. Zewail, Springer-Verlag, Berlin-Heidelberg, 1978, p. 258

44. Theory of Optical Dephasing in Condensed Phases.
K. E. Jones and A. H. Zewail
Springer Series in Chemical Physics, Vol. 3: Advances in Laser Chemistry, ed. A. H. Zewail, Springer-Verlag, Berlin-Heidelberg, 1978, p. 196

45. Optically Detected E.P.R. and Low-Field ENDOR of Triplet Benzophenone.
R. M. Hochstrasser, G. W. Scott, and A. H. Zewail
Mol. Phys. **36**, 475 (1978)

46. Photon Echoes and Coherence of Luminescent Molecules.
    A. H. Zewail and W. R. Lambert
    J. Lumin. **18–19**, 205 (1979)

47. Radiationless Relaxation and Optical Dephasing of Molecules
    Excited by Wide- and Narrow-Band Lasers. II. Pentacene in
    Low-Temperature Mixed Crystals.
    T. E. Orlowski and A. H. Zewail
    J. Chem. Phys. **70**, 1390 (1979)

48. Observation of High Energy Vibrational Overtones of
    Molecules in Solids: Local Modes and Intramolecular
    Relaxations.
    J. W. Perry and A. H. Zewail
    J. Chem. Phys. **70**, 582 (1979)

49. Solar Energy Flow Channels in an LSC.
    A. H. Zewail and J. S. Batchelder
    Proc. Photovoltaics Advanced Materials Review Meeting,
    DOE, Vail, 1979.

50. Are the Homogeneous Linewidths of Spin Resonance
    (ODMR) and Optical Transitions Related?
    A. H. Zewail
    J. Chem. Phys. **70**, 5759 (1979)

51. Solar Concentrators with Optimal Gain: LSC Development at
    Caltech.
    A. H. Zewail
    *Proc. Photovoltaic Concentrator Technology Development
    Project Integration Meeting,* DOE, Albuquerque, 1978

52. Energy Localization in Substitutionally Disordered Solids. II.
    Studies by Optical and Optically Detected Magnetic
    Resonance Spectroscopy.
    D. D. Smith, D. P. Millar, and A. H. Zewail
    J. Chem. Phys. **72**, 1187 (1980)

53. Vibronic Dephasing of Anharmonic Molecules. I. Theory and Its Application to the Separability of Intra- and Intermolecular Processes.
D. J. Diestler and A. H. Zewail
J. Chem. Phys. **71**, 3103 (1979)

54. Vibronic Dephasing of Anharmonic Molecules. II. Impurity Molecules Isolated in Low-Temperature Matrices.
D. J. Diestler and A. H. Zewail
J. Chem. Phys. **71**, 3113 (1979)

55. High-Energy Vibrational Overtones of Polyatomic Molecules in Solids: Inequivalent Local C-H Modes and Subpicosecond Dephasing of Naphthalene at 1.3 K.
J. W. Perry and A. H. Zewail
Chem. Phys. Lett. **65**, 31 (1979)

56. The Relaxation of Local Modes of Vibrationally-Hot Molecules in Cold Environments: Anharmonicity and Condensed Phase Effects.
A. H. Zewail and D. J. Diestler
Chem. Phys. Lett. **65**, 37 (1979)

57. Characterization of Vibrational Overtones and "Local" Modes by Emission Spectroscopy.
D. D. Smith and A. H. Zewail
J. Chem. Phys. **71**, 540 (1979)

58. Luminescent Solar Concentrators. I. Theory of Operation and Techniques for Performance Evaluation.
J. S. Batchelder, A. H. Zewail, and T. Cole
Appl. Opt. **18**, 3090 (1979)

59. Direct Measurement of Excitation Transport in a Highly Disordered Quasi-One-Dimensional Solid.
D. D. Smith, R. C. Powell, and A. H. Zewail
Chem. Phys. Lett. **68**, 309 (1979)

60. Spin-Quantization and Spin-Orbit Coupling Effects on the Line Shapes of Triplet States. II. The "Small" Exciton Problem.
J. P. Lemaistre and A. H. Zewail
J. Chem. Phys. **72**, 1055 (1980)

61. Picosecond Saturation Spectroscopy of Cresyl Violet: Rotational Diffusion by a "Sticking" Boundary Condition in the Liquid Phase.
D. P. Millar, R. R. Shah, and A. H. Zewail
Chem. Phys. Lett. **66**, 435 (1979)

62. Optical Observation of "Band-to-Band" Scattering by Time-Resolved Phosphorescence Line Narrowing: Exciton Dephasing in a Quasi-One-Dimensional Solid.
D. D. Smith and A. H. Zewail
J. Chem. Phys. **71**, 3533 (1979)

63. On the Inhomogeneous and Homogeneous Broadenings of Optical and ODMR Transitions of Triplet States in Solids.
J. P. Lemaistre and A. H. Zewail
Chem. Phys. Lett. **68**, 296 (1979)

64. Fluorescence, Phosphorescence, and ODMR Line Narrowing of Molecules in Solids.
J. P. Lemaistre and A. H. Zewail
Chem. Phys. Lett. **68**, 302 (1979)

65. Intersystem Crossing Rates of Pentacene in Naphthalene at 1.9 K Following Single-Mode Laser Excitation.
W. R. Lambert and A. H. Zewail
Chem. Phys. Lett. **69**, 270 (1980)

66. Optical Molecular Dephasing: Principles of and Probings by Coherent Laser Spectroscopy.
A. H. Zewail
Acc. Chem. Res. **13**, 360 (1980)

67. Dynamics of Molecular Excitons: Disorder, Coherence, and Dephasing.
A. H. Zewail, D. D. Smith, and J. P. Lemaistre
Modern Problems in Solid State Physics, Vol. 2: Excitons,
eds. E. I. Rashba and M. D. Sturge, North-Holland,
Amsterdam, 1982, p. 665

68. Laser Selective Chemistry: A New Challenge for Chemists and Physicists.
A. H. Zewail
Eng. Sci. **43**, 8 (1980)

69. Laser and Coherence Spectroscopy. Edited by Jeffrey I. Steinfeld.
Plenum, New York. 1978. pp. 530.
A. H. Zewail
J. Am. Chem. Soc. **102**, 7170 (1980)

70. Direct Observation of the Torsional Dynamics of DNA and RNA by Picosecond Spectroscopy.
D. P. Millar, R. J. Robbins, and A. H. Zewail
Proc. Natl. Acad. Sci. U.S.A. **77**, 5593 (1980)

71. Picosecond Torsional Dynamics of DNA.
R. J. Robbins, D. P. Millar, and A. H. Zewail
Springer Series in Chemical Physics, Vol. 14: Picosecond Phenomena II, eds. R. M. Hochstrasser, W. Kaiser, and
C. V. Shank, Springer-Verlag, Berlin-Heidelberg, 1980, p. 331

72. Observation of Dipolar-Induced Spin Dephasing in Ionic Solids Using Coherent Optical-Microwave Spectroscopy.
M. Glasbeek, R. Hond, and A. H. Zewail
Phys. Rev. Lett. **45**, 744 (1980)

73. Luminescent Solar Concentrators.
J. S. Batchelder, A. H. Zewail, and T. Cole
Proc. SPIE, Vol. 248: Role of Electro-Optics in Photovoltaic Energy Conversion, SPIE, Bellingham, 1980, p. 105

74. Exciton Dynamics in Quasi-One-Dimensional Molecular
    Systems.
    A. H. Zewail and D. D. Smith
    *Springer Series in Solid-State Sciences, Vol. 23: Physics in
    One Dimension*, eds. J. Bernasconi and T. Schneider,
    Springer-Verlag, Berlin-Heidelberg, 1981, p. 351

75. Laser Selective Chemistry: Is it Possible?
    A. H. Zewail
    Phys. Today **33**, 27 (1980)

76. Observation of Large Splittings, Narrow Resonances, and
    Polarization Extinction in the High-Energy Overtone Spectra
    of Large Molecules: Experimental Tests of Locality of
    Excitation in Bonds.
    J. W. Perry and A. H. Zewail
    J. Phys. Chem. **85**, 933 (1981)

77. Time-Resolved Spectroscopy of Macromolecules: Effect of
    Helical Structure on the Torsional Dynamics of DNA and RNA.
    D. P. Millar, R. J. Robbins, and A. H. Zewail
    J. Chem. Phys. **74**, 4200 (1981)

78. Observation of Intracavity Absorption of Molecules in
    Supersonic Beams.
    W. R. Lambert, P. M. Felker, and A. H. Zewail
    J. Chem. Phys. **74**, 4732 (1981)

79. Energy and Phase Randomization in Large Molecules as
    Probed by Laser Spectroscopy.
    A. H. Zewail
    Energy Storage and Redistribution in Molecules, ed.
    J. Hinze, Plenum, New York, 1983, p. 17

80. Picosecond Dynamics of Electronic Energy Transfer in
    Condensed Phases.
    D. P. Millar, R. J. Robbins, and A. H. Zewail
    J. Chem. Phys. **75**, 3649 (1981)

81. Optical Dephasing in Multilevel Systems: Bath-Independent Dephasing and Breakdown of the Two-Level Approximation.
W. S. Warren and A. H. Zewail
J. Phys. Chem. **85**, 2309 (1981)

82. Luminescent Solar Concentrators. II. Experimental and Theoretical Analysis of Their Possible Efficiencies.
J. S. Batchelder, A. H. Zewail, and T. Cole
Appl. Opt. **20**, 3733 (1981)

83. Laser-Selective Chemistry and Vibrational Energy Redistribution in Molecules.
A. H. Zewail
J. Photochem. **17**, 269 (1981)

84. Nonlinear Laser Spectroscopy and Dephasing of Molecules: An Experimental and Theoretical Overview.
M. J. Burns, W. K. Liu, and A. H. Zewail
Series in Modern Problems in Condensed Matter Sciences, Vol. 4: Spectroscopy and Excitation Dynamics of Condensed Molecular Systems, eds. V. M. Agranovich and R. M. Hochstrasser, North-Holland, Amsterdam, 1983, p. 301

85. Torsion and Bending of Nucleic Acids Studied by Subnanosecond Time-Resolved Fluorescence Depolarization of Intercalated Dyes.
D. P. Millar, R. J. Robbins, and A. H. Zewail
J. Chem. Phys. **76**, 2080 (1982)

86. Optical Analogues of NMR Phase Coherent Multiple Pulse Spectroscopy.
W. S. Warren and A. H. Zewail
J. Chem. Phys. **75**, 5956 (1981)

87. Quantum Beats and Dephasing in Isolated Large Molecules Cooled by Supersonic Jet Expansion and Excited by Picosecond Pulses: Anthracene.
W. R. Lambert, P. M. Felker, and A. H. Zewail
J. Chem. Phys. **75**, 5958 (1981)

88. Laser Probing of Vibrational Energy Redistribution and Dephasing.
    A. H. Zewail, W. R. Lambert, P. M. Felker, J. W. Perry, and W. S. Warren
    J. Phys. Chem. **86**, 1184 (1982)

89. Picosecond Excitation and *trans-cis* Isomerization of Stilbene in a Supersonic Jet: Dynamics and Spectra.
    J. A. Syage, W. R. Lambert, P. M. Felker, A. H. Zewail, and R. M. Hochstrasser
    Chem. Phys. Lett. **88**, 266 (1982)

90. Picosecond Excitation of Jet-Cooled Pyrazine: Magnetic Field Effects on the Fluorescence Decay and Quantum Beats.
    P. M. Felker, W. R. Lambert, and A. H. Zewail
    Chem. Phys. Lett. **89**, 309 (1982)

91. Luminescent Solar Concentrators: An Overview.
    A. H. Zewail and J. S. Batchelder
    ACS Symp. Ser., Vol. 220: Polymers in Solar Energy Utilization, eds. C. G. Gebelein, D. J. Williams, and R. D. Deanin, ACS, Washington, 1983, p. 331

92. Pulse Substructure in Passively and Synchronously Mode-Locke CW Dye Lasers.
    D. P. Millar and A. H. Zewail
    Chem. Phys. **72**, 381 (1982)

93. Picosecond Excitation of Jet-Cooled Hydrogen-Bonded Systems: Dispersed Fluorescence and Time-Resolved Studies of Methyl Salicylate.
    P. M. Felker, W. R. Lambert, and A. H. Zewail
    J. Chem. Phys. **77**, 1603 (1982)

94. Thermal Lensing Spectroscopy with Picosecond Pulse Trains and a New Dual Beam Configuration.
    J. W. Perry, E. A. Ryabov, and A. H. Zewail
    Laser Chem. **1**, 9 (1982)

95. Direct Observation of Intramolecular Energy Transfer by
    Selective Picosecond Laser Excitation of a Single
    Chromophore in Jet-Cooled Molecules.
    P. M. Felker, J. A. Syage, W. R. Lambert, and A. H. Zewail
    Chem. Phys. Lett. **92**, 1 (1982)

96. Local Modes: Their Relaxation, Polarization, and
    Stereoselective Excitation by Lasers.
    J. W. Perry and A. H. Zewail
    J. Phys. Chem. **86**, 5197 (1982)

97. Multiple Phase-Coherent Laser Pulses in Optical
    Spectroscopy. I. The Technique and Experimental Applications.
    W. S. Warren and A. H. Zewail
    J. Chem. Phys. **78**, 2279 (1983)

98. Multiple Phase-Coherent Laser Pulses in Optical Spectroscopy.
    II. Applications to Multilevel Systems.
    W. S. Warren and A. H. Zewail
    J. Chem. Phys. **78**, 2298 (1983)

99. Picosecond Laser Spectroscopy of Molecules in Supersonic
    Jets: Vibrational Energy Redistribution and Quantum Beats.
    A. H. Zewail
    *Springer Series in Chemical Physics, Vol. 23: Picosecond
    Phenomena III*, eds. K. B. Eisenthal, R. M. Hochstrasser,
    W. Kaiser, A. Laubereau, Springer-Verlag, Berlin-Heidelberg,
    1982, p. 184

100. Jet Spectroscopy of Isoquinoline.
     P. M. Felker and A. H. Zewail
     Chem. Phys. Lett. **94**, 448 (1983)

101. Stepwise Solvation of Molecules as Studied by Picosecond-
     Jet Spectroscopy: Dynamics and Spectra.
     P. M. Felker and A. H. Zewail
     Chem. Phys. Lett. **94**, 454 (1983)

102. Optical Multiple Pulse Sequences for Multiphoton Selective
     Excitation and Enhancement of Forbidden Transitions.
     W. S. Warren and A. H. Zewail
     J. Chem. Phys. **78**, 3583 (1983)

103. Photodissociation of Partially Solvated Molecules in Beams
     by the Picosecond-Jet Technique: Hydrogen Bond Breakage.
     P. M. Felker and A. H. Zewail
     J. Chem. Phys. **78**, 5266 (1983)

104. Picosecond-Jet Spectroscopy and Photochemistry: Energy
     Redistribution and Its Impact on Coherence, Isomerization,
     Dissociation, and Solvation.
     A. H. Zewail
     Faraday Discuss. Chem. Soc. **75**, 315 (1983); *100 Years of
     Physical Chemistry: A Celebration of the Faraday Society*,
     RSC, Cambridge, 2003, p. 107

105. Laser Spectroelectrochemistry of a Ruthenium (II) Tris
     (Bipyridyl) Derivative Adsorbed on Graphite Electrodes.
     J. W. Perry, A. J. McQuillan, F. C. Anson, and A. H. Zewail
     J. Phys. Chem. **87**, 1480 (1983)

106. Phase Coherence in Multiple Pulse Optical Spectroscopy.
     W. S. Warren and A. H. Zewail
     Laser Chem. **2**, 37 (1983); *Photochemistry and
     Photobiology*, ed. A. H. Zewail, Harwood Academic,
     London, 1983, p. 37

107. Energy and Phase Relaxation of Phosphorescent F Centers
     in CaO.
     M. Glasbeek, D. D. Smith, J. W. Perry, W. R. Lambert, and
     A. H. Zewail
     J. Chem. Phys. **79**, 2145 (1983)

108. Picosecond Laser Chemistry in Supersonic Jet Beams.
     A. H. Zewail
     Laser Chem. **2**, 55 (1983); *Photochemistry and Photobiology*,
     ed. A. H. Zewail, Harwood Academic, London, 1983, p. 55

109. Luminescent Solar Concentrators (LSC): Their Physics and Chemistry.
T. Cole and A. H. Zewail
Photochemistry and Photobiology, ed. A. H. Zewail, Harwood Academic, London, 1983, p. 743

110. Unimolecular Reactions at Low Energies and RRKM Behavior: Isomerization and Dissociation.
L. R. Khundkar, R. A. Marcus, and A. H. Zewail
J. Phys. Chem. **87**, 2473 (1983)

111. Laser-Selective Spectroscopy and Optical Dephasing of Phosphorescent F Centers.
M. Glasbeek and A. H. Zewail
Photochemistry and Photobiology, ed. A. H. Zewail, Harwood Academic, London, 1983, p. 1123

112. Observation of Restricted IVR in Large Molecules: Quasi-Periodic Behavior, Phase-Shifted and Non-Phase-Shifted Quantum Beats.
P. M. Felker and A. H. Zewail
Chem. Phys. Lett. **102**, 113 (1983)

113. Diphenylbutadiene in Supersonic Jets: Spectroscopy and Picosecond Dynamics.
J. F. Shepanski, B. W. Keelan, and A. H. Zewail
Chem. Phys. Lett. **103**, 9 (1983)

114. Jet-Cooled Styrene: Spectra and Isomerization.
J. A. Syage, F. Al Adel, and A. H. Zewail
Chem. Phys. Lett. **103**, 15 (1983)

115. Hole-Burning and Zero-Field ODMR of Quasi-Linear Chain Excitons in 1,4-Dibromonaphthalene.
M. Glasbeek, R. Sitters, J. H. Scheijde, and A. H. Zewail
Chem. Phys. Lett. **102**, 475 (1983)

116. Picosecond Pump-Probe Multiphoton Ionization of Isolated Molecules: IVR and Coherence.
J. W. Perry, N. F. Scherer, and A. H. Zewail
Chem. Phys. Lett. **103**, 1 (1983)

117. High-Energy Overtone Spectroscopy of Some Deuterated Methanes.
J. W. Perry, D. J. Moll, A. Kuppermann, and A. H. Zewail
J. Chem. Phys. **82**, 1195 (1985)

118. Jet Spectroscopy of Anthracene and Deuterated Anthracenes.
W. R. Lambert, P. M. Felker, J. A. Syage, and A. H. Zewail
J. Chem. Phys. **81**, 2195 (1984)

119. Picosecond Excitation and Selective Intramolecular Rates in Supersonic Molecular Beams. I. SVL Fluorescence Spectra and Lifetimes of Anthracene and Deuterated Anthracenes.
W. R. Lambert, P. M. Felker, and A. H. Zewail
J. Chem. Phys. **81**, 2209 (1984)

120. Picosecond Excitation and Selective Intramolecular Rates in Supersonic Molecular Beams. II. Intramolecular Quantum Beats and IVR.
W. R. Lambert, P. M. Felker, and A. H. Zewail
J. Chem. Phys. **81**, 2217 (1984)

121. Picosecond Photo-Chemistry and Spectroscopy in Supersonic Beams.
P. M. Felker and A. H. Zewail
Applications of Picosecond Spectroscopy to Chemistry,
ed. K. B. Eisenthal, D. Reidel, Dordrecht, 1984, p. 273

122. High Energy CH-Overtone Spectra of Benzene at 1.8 K.
J. W. Perry and A. H. Zewail
J. Chem. Phys. **80**, 5333 (1984)

123. Picosecond Excitation and Selective Intramolecular Rates in Supersonic Molecular Beams. III. Photochemistry and Rates of a Charge Transfer Reaction.
J. A. Syage, P. M. Felker, and A. H. Zewail
J. Chem. Phys. **81**, 2233 (1984)

124. Laser Spectroscopy in Molecular Beams: An Automated System for Obtaining High Resolution Spectra of Large Molecules.
B. W. Keelan, J. A. Syage, J. F. Shepanski, and A. H. Zewail
Proc. Int. Conf. Lasers '83, ed. R. C. Powell, STS Press, McLean, 1985, p. 718

125. Unimolecular Reaction Rates in Solution and in the Isolated Molecule: Comparison of Diphenyl Butadiene Nonradiative Decay in Solutions and Supersonic Jets.
S. H. Courtney, G. R. Fleming, L. R. Khundkar, and A. H. Zewail
J. Chem. Phys. **80**, 4559 (1984)

126. Picosecond Pump-Probe and Polarization Techniques in Supersonic Molecular Beams: Measurement of Ultrafast Vibrational-Rotational Dephasing and Coherence.
N. F. Scherer, J. F. Shepanski, and A. H. Zewail
J. Chem. Phys. **81**, 2181 (1984)

127. Picosecond Dynamics and Photoisomerization of Stilbene in Supersonic Beams. I. Spectra and Mode Assignments.
J. A. Syage, P. M. Felker, and A. H. Zewail
J. Chem. Phys. **81**, 4685 (1984)

128. Picosecond Dynamics and Photoisomerization of Stilbene in Supersonic Beams. II. Reaction Rates and Potential Energy Surface.
J. A. Syage, P. M. Felker, and A. H. Zewail
J. Chem. Phys. **81**, 4706 (1984)

129. Energy Redistribution in Isolated Molecules and the Question of Mode-Selective Laser Chemistry Revisited.
N. Bloembergen and A. H. Zewail
J. Phys. Chem. **88**, 5459 (1984)

130. Direct Observation of Nonchaotic Multilevel Vibrational Energy Flow in Isolated Polyatomic Molecules.
P. M. Felker and A. H. Zewail
Phys. Rev. Lett. **53**, 501 (1984)

131. Direct Picosecond Time Resolution of Dissipative Intramolecular Vibrational-Energy Redistribution (IVR) in Isolated Molecules.
P. M. Felker and A. H. Zewail
Chem. Phys. Lett. **108**, 303 (1984)

132. Picosecond Chemistry of Collisionless Molecules in Supersonic Beams.
A. H. Zewail
Springer Series in Chemical Physics, Vol. 38: Ultrafast Phenomena IV, eds. D. H. Auston and K. B. Eisenthal, Springer-Verlag, Berlin-Heidelberg, 1984, p. 284

133. Phase- and Energy-Changing Collisions in Iodine Gas: Studies by Optical Multiple-Pulse Spectroscopy.
E. T. Sleva and A. H. Zewail
Chem. Phys. Lett. **110**, 582 (1984)

134. The Highly Excited C-H Stretching States of CHD_3, CHT_3, and CH_3D.
G. A. Voth, R. A. Marcus, and A. H. Zewail
J. Chem. Phys. **81**, 5494 (1984)

135. Intramolecular Dephasing in Pyrazine: Direct Picosecond Time Resolution.
J. L. Knee, F. E. Doany, and A. H. Zewail
J. Chem. Phys. **82**, 1042 (1985)

136. Rotational Band Contour Analysis of Symmetries and
     Interactions of Vibrational Levels in Anthracene S_1.
     B. W. Keelan and A. H. Zewail
     J. Chem. Phys. **82**, 3011 (1985)

137. Mode-Specific Intramolecular Vibrational Energy
     Redistribution: Direct Picosecond Measurements.
     P. M. Felker and A. H. Zewail
     J. Phys. Chem. **88**, 6106 (1984)

138. Picosecond Excitation and Selective Intramolecular Rates in
     Supersonic Molecular Beams. IV. Alkylanthracenes.
     J. A. Syage, P. M. Felker, D. H. Semmes, F. Al Adel, and
     A. H. Zewail
     J. Chem. Phys. **82**, 2896 (1985)

139. Dynamics of Intramolecular Vibrational-Energy
     Redistribution (IVR). I. Coherence Effects.
     P. M. Felker and A. H. Zewail
     J. Chem. Phys. **82**, 2961 (1985)

140. Dynamics of Intramolecular Vibrational-Energy
     Redistribution (IVR). II. Excess Energy Dependence.
     P. M. Felker and A. H. Zewail
     J. Chem. Phys. **82**, 2975 (1985)

141. Dynamics of Intramolecular Vibrational-Energy
     Redistribution (IVR). III. Role of Molecular Rotations.
     P. M. Felker and A. H. Zewail
     J. Chem. Phys. **82**, 2994 (1985)

142. Dynamics of Intramolecular Vibrational-Energy
     Redistribution (IVR). IV. Excess Energy Dependence,
     t-Stilbene.
     P. M. Felker, W. R. Lambert, and A. H. Zewail
     J. Chem. Phys. **82**, 3003 (1985)

143. IVR: Its Coherent and Incoherent Dynamics.
A. H. Zewail
Ber. Bunsenges. Phys. Chem. **89**, 264 (1985)

144. Ultrafast Molecular Relaxation of Isolated Stilbene:
Measurements by Picosecond Pump-Probe Techniques.
N. F. Scherer, J. W. Perry, F. E. Doany, and A. H. Zewail
J. Phys. Chem. **89**, 894 (1985)

145. Picosecond Mass Spectrometry of a Collisionless
Photodissociation Reaction.
J. L. Knee, L. R. Khundkar, and A. H. Zewail
J. Chem. Phys. **82**, 4715 (1985)

146. Picosecond Monitoring of a Chemical Reaction in
Molecular Beams:
Photofragmentation of R-I → R‡ + I.
J. L. Knee, L. R. Khundkar, and A. H. Zewail
J. Chem. Phys. **83**, 1996 (1985)

147. Picosecond Fluorescence Studies on Intramolecular
Photochemical Electron Transfer in Porphyrins Linked to
Quinones at Two Different Fixed Distances.
B. A. Leland, A. D. Joran, P. M. Felker, J. J. Hopfield,
A. H. Zewail, and P. B. Dervan
J. Phys. Chem. **89**, 5571 (1985)

148. IVR in Isolated Molecules with Nearby Electronic States.
J. L. Knee, L. R. Khundkar, and A. H. Zewail
J. Phys. Chem. **89**, 3201 (1985)

149. Rotational Band Contour Studies of Single Vibronic Levels
in Jet-Cooled *trans*-Stilbene.
B. W. Keelan and A. H. Zewail
J. Phys. Chem. **89**, 4939 (1985)

150. Rates of Photoisomerization of *trans*-Stilbene in Isolated
and Solvated Molecules: Experiments on the Deuterium

Isotope Effect and RRKM Behavior.
P. M. Felker and A. H. Zewail
J. Phys. Chem. **89**, 5402 (1985)

151. Practical Broad-Band Tuning of Dye Lasers by Solvent Shifting.
S. M. Grenci, G. R. Bird, B. W. Keelan, and A. H. Zewail
Laser Chem. **6**, 361 (1986)

152. Picosecond Photofragmentation of Tri- and Tetraatomic Molecules: Measurement of "State-to-State" Reaction Rates.
J. L. Knee, L. R. Khundkar, and A. H. Zewail
J. Phys. Chem. **89**, 4659 (1985)

153. Photon Locking.
E. T. Sleva, I. M. Xavier, Jr., and A. H. Zewail
J. Opt. Soc. Am. B **3**, 483 (1986)

154. Exciton and Vibronic Effects in the Spectroscopy of Bianthracene in Supersonic Beams.
L. R. Khundkar and A. H. Zewail
J. Chem. Phys. **84**, 1302 (1986)

E.1 *Excerpt from Dr. Zewail's autobiography*

"My science family came from all over the world, and members were of varied backgrounds, cultures, and abilities. The diversity in this 'small world' I worked in daily provided the most stimulating environment, with many challenges and much optimism. Over the years, my research group has had close to 150 graduate students, postdoctoral fellows, and visiting associates. Many of them are now in leading academic, industrial and governmental positions. Working with such minds in a village of science has been the most rewarding experience — Caltech was the right place for me."

E.2 *Dr. Zewail's publications during the beginning stage of his femtochemistry works in Caltech*

155. Femtosecond Photofragment Spectroscopy: The Reaction
ICN → CN + I.
N. F. Scherer, J. L. Knee, D. D. Smith, and A. H. Zewail
J. Phys. Chem. **89**, 5141 (1985)

156. Direct Picosecond Time Resolution of Unimolecular
Reactions Initiated by Local Mode Excitation.
N. F. Scherer, F. E. Doany, A. H. Zewail, and J. W. Perry
J. Chem. Phys. **84**, 1932 (1986)

157. Doppler-Free Time-Resolved Polarization Spectroscopy of
Large Molecules: Measurement of Excited State Rotational
Constants.
J. S. Baskin, P. M. Felker, and A. H. Zewail
J. Chem. Phys. **84**, 4708 (1986)

158. Rephasing of Collisionless Molecular Rotational
Coherence in Large Molecules.
P. M. Felker, J. S. Baskin, and A. H. Zewail
J. Phys. Chem. **90**, 724 (1986)

159. Locking of Dephasing and Energy Redistribution in
Molecular Systems by Multiple-Pulse Laser Excitation.
E. T. Sleva, M. Glasbeek, and A. H. Zewail
J. Phys. Chem. **90**, 1232 (1986)

160. Exciton Coherence in Highly Disordered Solid DBN:
"Band-to-Band" ODMR Transitions in Zero Field.
P. M. H. L. Tegelarr, M. Glasbeek, and A. H. Zewail
Chem. Phys. Lett. **128**, 455 (1986)

161. Phase Locking of Molecular Two-Level Quantum Systems:
Application to Ionic Solids.
R. Vreeker, M. Glasbeek, E. T. Sleva, and A. H. Zewail
Chem. Phys. Lett. **129**, 117 (1986)

162. Intramolecular Dephasing in Pyrazine: Deuterium Isotope
     Effect and Further Tests of Theory.
     P. M. Felker and A. H. Zewail
     Chem. Phys. Lett. **128**, 221 (1986)

163. Chemical Reaction Dynamics and Marcus' Contributions.
     R. B. Bernstein and A. H. Zewail
     J. Phys. Chem. **90**, 3467 (1986)

164. Picosecond and Femtosecond Molecular Beam Chemistry:
     Coherence and Fragment Recoil Dynamics.
     A. H. Zewail
     *Springer Series in Chemical Physics, Vol. 46: Ultrafast
     Phenomena V*, eds. G. R. Fleming and A. E. Siegman,
     Springer-Verlag, Berlin-Heidelberg, 1986, p. 356

165. Real-Time Measurements of IVR Versus Inferences from
     Spectral Broadening Data: The Alkylanilines "Ring + Tail"
     System.
     J. S. Baskin, M. Dantus, and A. H. Zewail
     Chem. Phys. Lett. **130**, 473 (1986)

166. Coherent Photodissociation Reactions: Observation by a
     Novel Picosecond Polarization Technique.
     J. S. Baskin, D. Semmes, and A. H. Zewail
     J. Chem. Phys. **85**, 7488 (1986)

167. Purely Rotational Coherence Effect and Time-Resolved
     Sub-Doppler Spectroscopy of Large Molecules. I. Theoretical.
     P. M. Felker and A. H. Zewail
     J. Chem. Phys. **86**, 2460 (1987)

168. Purely Rotational Coherence Effect and Time-Resolved
     Sub-Doppler Spectroscopy of Large Molecules. II.
     Experimental.
     J. S. Baskin, P. M. Felker, and A. H. Zewail
     J. Chem. Phys. **86**, 2483 (1987)

169. Picosecond Time-Resolved Dynamics of Vibrational-Energy Redistribution and Coherence in Beam-Isolated Molecules.
P. M. Felker and A. H. Zewail
Advances in Chemical Physics, Vol. 70: Evolution of Size Effects in Chemical Dynamics, Part 1, eds. I. Prigogine and S. A. Rice, John Wiley & Sons, New York, 1988, p. 265

170. Picosecond Photofragment Spectroscopy. I. Microcanonical State-to-State Rates of the Reaction NCNO $\rightarrow$ CN + NO.
L. R. Khundkar, J. L. Knee, and A. H. Zewail
J. Chem. Phys. **87**, 77 (1987)

171. Picosecond Photofragment Spectroscopy. II. The Overtone Initiated Unimolecular Reaction H_2O_2 (vOH= 5) $\rightarrow$ 2OH.
N. F. Scherer and A. H. Zewail
J. Chem. Phys. **87**, 97 (1987)

172. Picosecond Photofragment Spectroscopy. III. Vibrational Predissociation of van der Waals' Clusters.
J. L. Knee, L. R. Khundkar, and A. H. Zewail
J. Chem. Phys. **87**, 115 (1987)

173. Direct Observation of a Mode-Selective (Non-RRKM) van der Waals' Reaction by Picosecond Photofragment Spectroscopy.
D. H. Semmes, J. S. Baskin, and A. H. Zewail
J. Am. Chem. Soc. **109**, 4104 (1987)

174. Effect of Exothermicity on Electron Transfer Rates in Photosynthetic Molecular Models.
A. D. Joran, B. A. Leland, P. M. Felker, A. H. Zewail, J. J. Hopfield, and P. B. Dervan
Nature **327**, 508 (1987)

175. Real-Time Picosecond Clocking of the Collision Complex in a Bimolecular Reaction: The Birth of OH from H + CO_2.

N. F. Scherer, L. R. Khundkar, R. B. Bernstein, and A. H. Zewail
J. Chem. Phys. **87**, 1451 (1987)

176. Stepwise Solvation of the Intramolecular-Charge-Transfer Molecule, *p*-(Dimethylamino) Benzonitrile.
L. W. Peng, M. Dantus, A. H. Zewail, K. Kemnitz, J. M. Hicks, and K. B. Eisenthal
J. Phys. Chem. **91**, 6162 (1987)

E.3 *Excerpt from Dr. Zewail's autobiography*

"The journey from Egypt to America has been full of surprises. As a Moeid, I was unaware of the Nobel Prize but I now see its impact in the West. We used to gather around the TV or read in the newspaper about the recognition of famous Egyptian scientists and writers by the president, and these moments gave me and my friends a real thrill — maybe one day we would be in this position ourselves as a result of our achievements in science or literature."

E.4 *Dr. Zewail's publication on his brilliant research works in Caltech*

177. Real-Time Femtosecond Probing of "Transition States" in Chemical Reactions.
M. Dantus, M. J. Rosker, and A. H. Zewail
J. Chem. Phys. **87**, 2395 (1987)

178. Time-Resolved Spin Diffusion of Quasi-Linear Excitons in Disordered Solid DBN.
P. M. H. L. Tegelaar, M. Glasbeek, and A. H. Zewail
Chem. Phys. Lett. **140**, 1 (1987)

179. Synchrotron Radiation Selected *s*-Tetrazine Ion Chemistry.
I. Nenner, O. Dutuit, M. Richard-Viard, P. Morin, and A. H. Zewail
J. Am. Chem. Soc. **110**, 1093 (1988)

180. Sub-Doppler Measurement of Excited-State Rotational Constants and Rotational Coherence by Picosecond Multiphoton Ionization Mass Spectrometry.
N. F. Scherer, L. R. Khundkar, T. S. Rose, and A. H. Zewail
J. Phys. Chem. **91**, 6478 (1987)

181. Picosecond IVR Dynamics of *p*-Difluorobenzene and *p*-Fluorotoluene in a Molecular Beam: Comparison with Chemical Timing Data.
J. S. Baskin, T. S. Rose, and A. H. Zewail
J. Chem. Phys. **88**, 1458 (1988)

182. Time Dependent Absorption of Fragments During Dissociation.
R. Bersohn and A. H. Zewail
Ber. Bunsenges. Phys. Chem. **92**, 373 (1988)

183. Picosecond MPI Mass Spectrometry of CH_3I in the Process of Dissociation.
L. R. Khundkar and A. H. Zewail
Chem. Phys. Lett. **142**, 426 (1987)

184. Real-Time Femtochemistry.
A. H. Zewail and R. B. Bernstein
Kagaku to Kogyo (Science and Industry) **41**, 298 (1988)

185. Femtosecond Real-Time Observation of Wave Packet Oscillations (Resonance) in Dissociation Reactions.
T. S. Rose, M. J. Rosker, and A. H. Zewail
J. Chem. Phys. **88**, 6672 (1988)

186. Dynamics of Intramolecular Vibrational Energy Redistribution in Deuteriated Anthracenes: Rotational Band Contour Analysis and Time-Resolved Measurements.
L. W. Peng, B. W. Keelan, D. H. Semmes, and A. H. Zewail
J. Phys. Chem. **92**, 5540 (1988)

187. Ultrafast Laser Spectroscopy of Chemical Reactions.
J. L. Knee and A. H. Zewail
Spectroscopy **3**, 44 (1988)

188. Real-Time Laser Femtochemistry: Viewing the Transition
States from Reagents to Products.
A. H. Zewail and R. B. Bernstein
Chem. Eng. News, **66**, 24 (1988);
The Chemical Bond: Structure and Dynamics, ed. A. H.
Zewail, Academic Press, Boston, 1992, p. 223

189. Femtosecond Real-Time Dynamics of Photofragment-
Trapping Resonances on Dissociative Potential Energy
Surfaces.
M. J. Rosker, T. S. Rose, and A. H. Zewail
Chem. Phys. Lett. **146**, 175 (1988)

190. Application of Unimolecular Reaction Rate Theory for
Highly Flexible Transition States to the Dissociation of
NCNO into NC and NO.
S. J. Klippenstein, L. R. Khundkar, A. H. Zewail, and
R. A. Marcus
J. Chem. Phys. **89**, 4761 (1988)

191. Femtosecond Clocking of the Chemical Bond.
M. J. Rosker, M. Dantus, and A. H. Zewail
Science **241**, 1200 (1988)

192. Photon Locking and Its Observation by the Probe-Echo
Method: Application to Optical and Microwave Transitions.
R. Vreeker, M. Glasbeek, and A. H. Zewail
J. Phys. Chem. **93**, 658 (1989)

193. Femtosecond Spectroscopy of Transition States in
Reactions.
A. H. Zewail
*Springer Series in Chemical Physics, Vol. 48: Ultrafast
Phenomena VI*, eds. T. Yajima, K. Yoshihara, C. B. Harris,

and S. Shionoya, Springer-Verlag, Berlin-Heidelberg, 1988, p. 500

194. Femtosecond Real-Time Probing of Reactions. I. The Technique.
M. J. Rosker, M. Dantus, and A. H. Zewail
J. Chem. Phys. **89**, 6113 (1988)

195. Femtosecond Real-Time Probing of Reactions. II. The Dissociation Reaction of ICN.
M. Dantus, M. J. Rosker, and A. H. Zewail
J. Chem. Phys. **89**, 6128 (1988)

196. Determination of Excited-State Rotational Constants and Structures by Doppler-Free Picosecond Spectroscopy.
J. S. Baskin and A. H. Zewail
J. Phys. Chem. **93**, 5701 (1989)

197. Femtosecond Real-Time Probing of Reactions. III. Inversion to the Potential from Femtosecond Transition-State Spectroscopy Experiments.
R. B. Bernstein and A. H. Zewail
J. Chem. Phys. **90**, 829 (1989)

198. Laser Femtochemistry.
A. H. Zewail
Science **242**, 1645 (1988)

199. Femtochemistry: The Role of Alignment and Orientation.
A. H. Zewail
J. Chem. Soc. Faraday Trans. **285**, 1221 (1989)

200. Picosecond Studies of Jet-Cooled Chromyl Chloride.
D. S. Tinti, J. S. Baskin, and A. H. Zewail
Chem. Phys. Lett. **155**, 243 (1989)

201. Molecular State Evolution After Excitation with an Ultra-Short Laser Pulse: A Quantum Analysis of NaI and NaBr Dissociation.

V. Engel, H. Metiu, R. Almeida, R. A. Marcus, and A. H. Zewail
Chem. Phys. Lett. **152**, 1 (1988)

202. Femtochemistry of the Reaction: $IHgI^* \rightarrow [IHg...I]\ddagger^* \rightarrow HgI + I$.
R. M. Bowman, M. Dantus, and A. H. Zewail
Chem. Phys. Lett. **156**, 131 (1989)

203. Femtosecond Real-Time Alignment in Chemical Reactions.
M. Dantus, R. M. Bowman, J. S. Baskin, and A. H. Zewail
Chem. Phys. Lett. **159**, 406 (1989)

204. Picosecond Photofragment Spectroscopy. IV. Dynamics of Consecutive Bond Breakage in the Reaction
$C_2F_4I_2 \rightarrow C_2F_4 + 2I$.
L. R. Khundkar and A. H. Zewail
J. Chem. Phys. **92**, 231 (1990)

205. Femtosecond Real-Time Probing of Reactions. IV. The Reactions of Alkali Halides.
T. S. Rose, M. J. Rosker, and A. H. Zewail
J. Chem. Phys. **91**, 7415 (1989)

206. Femtosecond Real-Time Probing of Reactions. V. The Reaction of IHgI.
M. Dantus, R. M. Bowman, M. Gruebele, and A. H. Zewail
J. Chem. Phys. **91**, 7437 (1989)

207. Femtosecond Transition-State Spectroscopy of Iodine: From Strongly Bound to Repulsive Surface Dynamics.
R. M. Bowman, M. Dantus, and A. H. Zewail
Chem. Phys. Lett. **161**, 297 (1989)

208. Real-Time Probing of Reactions in Clusters.
J. J. Breen, L. W. Peng, D. M. Willberg, A. A. Heikal,
P. Cong, and A. H. Zewail
J. Chem. Phys. **92**, 805 (1990)

209. Picosecond Dissociation of Ketene: Experimental State-to-State Rates and Tests of Statistical Theories.
E. D. Potter, M. Gruebele, L. R. Khundkar, and A. H. Zewail
Chem. Phys. Lett. **164**, 463 (1989)

210. Picosecond Real-Time Studies of Mode-Specific Vibrational Predissociation.
D. H. Semmes, J. S. Baskin, and A. H. Zewail
J. Chem. Phys. **92**, 3359 (1990)

211. Real-Time Clocking of Bimolecular Reactions: Application to $H + CO_2$.
N. F. Scherer, C. Sipes, R. B. Bernstein, and A. H. Zewail
J. Chem. Phys. **92**, 5239 (1990)

212. Ultrafast Molecular Reaction Dynamics in Real-Time: Progress Over a Decade.
L. R. Khundkar and A. H. Zewail
Ann. Rev. Phys. Chem. **41**, 15 (1990)

213. Femtosecond Laser Observations of Molecular Vibration and Rotation.
M. Dantus, R. M. Bowman, and A. H. Zewail
Nature **343**, 737 (1990)

214. Femtosecond Temporal Spectroscopy and Direct Inversion to the Potential: Application to Iodine.
M. Gruebele, G. Roberts, M. Dantus, R. M. Bowman, and A. H. Zewail
Chem. Phys. Lett. **166**, 459 (1990)

215. Femtochemistry of the Reaction of IHgI: Theory Versus Experiment.
M. Gruebele, G. Roberts, and A. H. Zewail
Phil. Trans. R. Soc. Lond. A **332**, 223 (1990)

216. Ultrafast Reaction Dynamics.
M. Gruebele and A. H. Zewail
Phys. Today **43**, 24 (1990); *Ber. Bunsenges. Phys. Chem.*
(German) **94**, 1210 (1990);
Parity (Japanese) **5**, 8 (1990);
Usp. Fiz. Nauk (Russian) **161**, 69 (1991)

217. From Femtosecond Temporal Spectroscopy to the Potential by a Direct Classical Inversion Method.
R. B. Bernstein and A. H. Zewail
Chem. Phys. Lett. **170**, 321 (1990)

218. Femtosecond Probing of Persistent Wave Packet Motion in Dissociative Reactions: Up to 40 Picoseconds.
P. Cong, A. Mokhtari, and A. H. Zewail
Chem. Phys. Lett. **172**, 109 (1990)

219. Femtosecond Temporal Spectroscopy of ICl: Inversion to the $A_3\Pi_1$ State Potential.
M. H. M. Janssen, R. M. Bowman, and A. H. Zewail
Chem. Phys. Lett. **172**, 99 (1990)

220. Femtosecond Selective Control of Wave Packet Population.
J. J. Gerdy, M. Dantus, R. M. Bowman, and A. H. Zewail
Chem. Phys. Lett. **171**, 1 (1990)

221. Lasers and Ultrafast Phenomena.
A. H. Zewail
Lasers in Science and Technology, eds. G. Eden and M. Nayfeh, Gordon & Breach, New York, 1990

222. The Birth of Molecules.
A. H. Zewail
Sci. Am. **263**, 76 (1990);
Le Scienze (Italian) **270**, 52 (1991);
Pour la Science (French) **160**, 92 (1991);
Investigacion y Ciencia (Spanish) **173**, 42, (1991);
Spektrum der Wissenschaft (German), February 1991, p. 100;
V Mire Hauki (Russian) **2**, 30 (1991);
Ke Xue (Chinese) **152**, 22 (1991);
Majallat Al Oloom (Arabic) **8**, 60 (1992);
Tudomány (Hungarian), February 1991, p. 28;
Vigyan (Indian);
Al Khairia (English);
Saiensu (Japanese);
Proceedings of the Royal Institution of Great Britain, Vol.
63, ed. S. T. Nash, Science Reviews, Northwood, 1991, p. 269

223. Direct Observation of the Picosecond Dynamics of I_2 – Ar Fragmentation.
J. J. Breen, D. M. Willberg, M. Gutmann, and A. H. Zewail
J. Chem. Phys. **93**, 9180 (1990)

224. Femtosecond Multiphoton Dynamics of Higher-Energy Potentials.
R. M. Bowman, M. Dantus, and A. H. Zewail
Chem. Phys. Lett. **174**, 546 (1990)

225. Direct Femtosecond Mapping of Trajectories in a Chemical Reaction.
A. Mokhtari, P. Cong, J. L. Herek, and A. H. Zewail
Nature **348**, 225 (1990)

226. Femtosecond Real-Time Probing of Reactions. VI. A Joint Experimental and Theoretical Study of Bi_2 Dissociation.
R. M. Bowman, J. J. Gerdy, G. Roberts, and A. H. Zewail
J. Phys. Chem. **95**, 4635 (1991)

227. Femtosecond Transition-State Dynamics.
A. H. Zewail
Faraday Discuss. Chem. Soc. **91**, 207 (1991)

228. Structural Femtochemistry: Experimental Methodology.
J. C. Williamson and A. H. Zewail
Proc. Natl. Acad. Sci. U.S.A. **88**, 5021 (1991)

229. Femtosecond Real-Time Probing of Reactions. VII. A
Quantum and Classical Mechanical Study of the ICN
Dissociation Experiment.
G. Roberts and A. H. Zewail
J. Phys. Chem. **95**, 7973 (1991)

230. Optical Receiver Based on Luminescent Light Trapping.
J. W. Perry, T. Cole, and A. H. Zewail
NASA Tech Briefs **15**, 59 (1991)

231. Real-Time Dynamics of Vibrational Predissociation in
Anthracene-Ar_n($n = 1,2,3$).
A. A. Heikal, L. Bañares, D. H. Semmes, and A. H. Zewail
Chem. Phys. **156**, 231 (1991)

232. Femtosecond Probing of Molecular Dynamics by Mass-
Spectrometry in a Molecular Beam.
M. Dantus, M. H. M. Janssen, and A. H. Zewail
Chem. Phys. Lett. **181**, 281 (1991)

233. Femtosecond Probing of Bimolecular Reactions: The
Collision Complex.
M. Gruebele, I. R. Sims, E. D. Potter, and A. H. Zewail
J. Chem. Phys. **95**, 7763 (1991)

234. Rotational Constants of Vibrationally Excited Iodine from
Purely Rotational Coherence Observed in Pump-Probe
Experiments.
D. M. Willberg, J. J. Breen, M. Gutmann, and A. H. Zewail
J. Phys. Chem. **95**, 7136 (1991)

235. Real-Time Dynamics of Clusters. I. $I_2X_n(n = 1)$.
D. M. Willberg, M. Gutmann, J. J. Breen, and A. H. Zewail
J. Chem. Phys. **96**, 198 (1992)

236. Femtochemistry: Recent Advances and Extension to High Pressures.
A. H. Zewail, M. Dantus, R. M. Bowman, and A. Mokhtari
J. Photochem. Photobiol. A **62**, 301 (1992)

237. Femtosecond Laser Control of a Chemical Reaction.
E. D. Potter, J. L. Herek, S. Pedersen, Q. Liu, and A. H. Zewail
Nature **355**, 66 (1992)

238. Femtosecond Real-Time Probing of Reactions. VIII. The Bimolecular Reaction of $Br + I_2$.
I. R. Sims, M. Gruebele, E. D. Potter, and A. H. Zewail
J. Chem. Phys. **97**, 4127 (1992)

239. Ultrafast Dynamics of Isomerization Reactions: Structural Effects in Stilbene(s).
L. Bañares, A. A. Heikal, and A. H. Zewail
J. Phys. Chem. **96**, 4127 (1992)

240. Femtosecond Chemical Dynamics in Solution: Wavepacket Evolution and Caging of I_2.
Y. Yan, R. M. Whitnell, K. R. Wilson, and A. H. Zewail
Chem. Phys. Lett. **193**, 402 (1992)

241. Real-Time Dynamics of Clusters. II. $I_2X_n(n = 1; X = He, Ne,$ and $H_2)$, Picosecond Fragmentation.
M. Gutmann, D. M. Willberg, and A. H. Zewail
J. Chem. Phys. **97**, 8037 (1992)

242. Real-Time Dynamics of Clusters. III. $I_2Ne_n(n = 2–4)$, Picosecond Fragmentation and Evaporation.
M. Gutmann, D. M. Willberg, and A. H. Zewail
J. Chem. Phys. **97**, 8048 (1992)

243. Femtosecond Wave Packet Spectroscopy: Coherences, the Potential, and Structural Determination.
M. Gruebele and A. H. Zewail
J. Chem. Phys. **98**, 883 (1993)

244. Femtochemistry.
A. H. Zewail
Springer Series in Chemical Physics, Vol. 55: Ultrafast Phenomena VIII, eds. J.-L. Martin, A. Migus, G. A. Mourou, and A. H. Zewail, Springer-Verlag, Berlin-Heidelberg, 1993, p. 43

245. Femtosecond Real-Time Probing of Reactions. IX. Hydrogen Atom Transfer.
J. L. Herek, S. Pedersen, L. Bañares, and A. H. Zewail
J. Chem. Phys. **97**, 9046 (1992)

246. Ultrafast Diffraction and Molecular Structure.
J. C. Williamson, M. Dantus, S. B. Kim, and A. H. Zewail
Chem. Phys. Lett. **196**, 529 (1992)

247. Femtosecond Vibrational Transition-State Dynamics in a Chemical Reaction.
S. Pedersen, L. Bañares, and A. H. Zewail
J. Chem. Phys. **97**, 8801 (1992)

248. Ultrafast Dynamics in Solution: Wavepacket Motion and the Cage Effect in Iodine.
Y. Yan, R. M. Whitnell, K. R. Wilson, and A. H. Zewail
Springer Series in Chemical Physics, Vol. 55: Ultrafast Phenomena VIII, eds. J.-L. Martin, A. Migus, G. A. Mourou, and A. H. Zewail,
Springer-Verlag, Berlin-Heidelberg, 1993, p. 74

249. Chemistry in Femtotime.
A. H. Zewail
Vacuum Ultraviolet Radiation Physics, eds. F. J. Wuilleumier, Y. Petroff, and I. Nenner, World Scientific, Singapore, 1993, p. 20

250. A Simple Description of Vibrational Predissociation by a
     Full-Collision Approach.
     D. M. Willberg, M. Gutmann, E. E. Nikitin, and A. H. Zewail
     Chem. Phys. Lett. **201**, 506 (1993)

251. Femtosecond Reaction Dynamics in Macroclusters: Effect
     of Solvation on Wave Packet Motion.
     E. D. Potter, Q. Liu, and A. H. Zewail
     Chem. Phys. Lett. **200**, 605 (1992)

252. Femtosecond Real-Time Probing of Reactions: X. Reaction
     Times and Model Potentials.
     Q. Liu and A. H. Zewail
     J. Phys. Chem. **97**, 2209 (1993)

253. Small is Beautiful ("Taming the Atom" by Hans Christian
     von Baeyer, Random House, New York, 1992, pp. 223).
     A. H. Zewail
     Nature **361**, 215 (1993)

Fig. 11.2   Dr. Zewail receiving the King Faisal Prize (1989).

254. Ultrafast Lasers and Chemistry at Femtosecond Resolution.
A. H. Zewail
Science and Medicine in the 21st Century: A Global Perspective, The King Faisal Foundation, Riyadh-London, p. 227, 1994

255. Ultrafast Electron Diffraction. Velocity Mismatch and Temporal Resolution in Cross-Beam Experiments.
J. C. Williamson and A. H. Zewail
Chem. Phys. Lett. **209**, 10 (1993)

256. Femtosecond Real-Time Probing of Reactions. XI. The Elementary OClO Fragmentation.
T. Baumert, J. L. Herek, and A. H. Zewail
J. Chem. Phys. **99**, 4430 (1993)

257. Femtosecond Dynamics of Dissociation and Recombination in Solvent Cages.
Q. Liu, J.-K. Wang, and A. H. Zewail
Nature **364**, 427 (1993)

258. Picosecond Dynamics of *n*-Hexane Solvated *trans*-Stilbene.
C. Lienau, A. A. Heikal, and A. H. Zewail
Chem. Phys. **175**, 171 (1993)

259. Femtosecond Reaction Dynamics of Rydberg States: Methyl Iodide.
M. H. M. Janssen, M. Dantus, H. Guo, and A. H. Zewail
Chem. Phys. Lett. **214**, 281 (1993)

260. Femtosecond Reaction Dynamics: Elementary Processes, from Isolated to Solvated Reactions.
A. H. Zewail
Femtosecond Reaction Dynamics, ed. D. A. Wiersma, North Holland, Amsterdam, 1994, p. 1

261. Femtochemistry at High Pressures: The Dynamics of an
     Elementary Reaction in the Gas-Liquid Transition Region.
     C. Lienau, J. C. Williamson, and A. H. Zewail
     Chem. Phys. Lett. **213**, 289 (1993)

262. Femtochemistry.
     A. H. Zewail
     J. Phys. Chem. **97**, 12427 (1993)

263. Femtosecond Real-Time Probing of Reactions. XII.
     Vectorial Dynamics of Transition States.
     T. Baumert, S. Pedersen, and A. H. Zewail
     J. Phys. Chem. **97**, 12447 (1993)

264. Femtosecond Real-Time Probing of Reactions. XIII.
     Multiphoton Dynamics of IHgI.
     S. Pedersen, T. Baumert, and A. H. Zewail
     J. Phys. Chem. **97**, 12460 (1993)
     1st. INT. Conf. on LASERS & Applications, NILES, Cairo
     Univ. March (1994)

265. Femtosecond Real-Time Probing of Reactions. XIV.
     Rydberg States of Methyl Iodide.
     H. Guo and A. H. Zewail
     Can. J. Chem. **72**, 947 (1994)

266. Femtochemistry: Concepts and Applications.
     A. H. Zewail
     Femtosecond Chemistry, eds. J. Manz and L. Wöste, VCH,
     Weinheim, 1995, p. 15

267. Molecular Structures from Ultrafast Coherence Spectroscopy.
     P. M. Felker and A. H. Zewail
     Femtosecond Chemistry, eds. J. Manz and L. Wöste, VCH,
     Weinheim, 1995, p. 193

268. Ultrafast Dynamics of the Chemical Bond: Femtochemistry.
     A. H. Zewail
     Ultrafast Processes in Chemistry and Photobiology, eds.

M. A. El-Sayed, I. Tanaka, and Y. Molin, University Press, Cambridge, 1995, p. 1

269. Femtosecond Real-Time Probing of Reactions. XV. Time-Dependent Coherent Alignment.
J. S. Baskin and A. H. Zewail
J. Phys. Chem. **98**, 3337 (1994)

270. Femtosecond Real-Time Probing of Reactions. XVI. Dissociation with Intense Pulses.
A. Materny, J. L. Herek, P. Cong, and A. H. Zewail
J. Phys. Chem. **98**, 3352 (1994)

271. Ultrafast Electron Diffraction. IV. Molecular Structures and Coherent Dynamics.
J. C. Williamson and A. H. Zewail
J. Phys. Chem. **98**, 2766 (1994)

272. Ultrafast Electron Diffraction. V. Experimental Time Resolution and Applications.
M. Dantus, S. B. Kim, J. C. Williamson, and A. H. Zewail
J. Phys. Chem. **98**, 2782 (1994)

273. Rotational Coherence Phenomena.
P. M. Felker and A. H. Zewail
Jet Spectroscopy and Molecular Dynamics, eds. J. M. Hollas and D. Phillips, Chapman & Hall, Blackie Academic, London, 1995, p. 181

274. Ultrafast Dynamics of IVR in Molecules and Reactions.
P. M. Felker and A. H. Zewail
Jet Spectroscopy and Molecular Dynamics, eds. J. M. Hollas and D. Phillips, Chapman & Hall, Blackie Academic, London, 1995, p. 222

275. Femtochemistry at High Pressures: Solvent Effect in the Gas-to-Liquid Transition Region.
C. Lienau and A. H. Zewail
Chem. Phys. Lett. **222**, 224 (1994)

276. Femtosecond Control of an Elementary Unimolecular Reaction From the Transition-State Region.
J. L. Herek, A. Materny, and A. H. Zewail
Chem. Phys. Lett. **228**, 15 (1994)

277. Femtosecond pH jump: Dynamics of Acid-Base Reactions in Solvent Cages.
S. K. Kim, J.-K. Wang, and A. H. Zewail
Chem. Phys. Lett. **228**, 369 (1994)

278. Solvation Ultrafast Dynamics of Reactions. VIII. Acid-Base Reactions in Finite-Sized Clusters of Naphthol in Ammonia, Water, and Piperidine.
S. K. Kim, J. J. Breen, D. M. Willberg, L. W. Peng, A. A. Heikal, J. A. Syage, and A. H. Zewail
J. Phys. Chem. **99**, 7421 (1995)

279. The Validity of the "Diradical" Hypothesis: Direct Femtosecond Studies of the Transition-State Structures.
S. Pedersen, J. L. Herek, and A. H. Zewail
Science **266**, 1359 (1994)

280. Solvation Ultrafast Dynamics of Reactions. V. Dissociation and Atom Recombination of Iodine in the Gas-to-Liquid Transition Region.
C. Lienau and A. H. Zewail
J. Chim. Phys. **92**, 566 (1995)

281. Femtosecond Real-Time Probing of Reactions. XVII. Centrifugal Effects in Direct Dissociation Reactions.
G. Roberts and A. H. Zewail
J. Phys. Chem. **99**, 2520 (1995)

282. Transient Species at Femtosecond Resolution.
A. H. Zewail
Proc. 38th Robert A. Welch Conf. on Chemical Research: Chemical Dynamics of Transient Species, Welch Foundation, Houston, 1994, p. 129

283. Dynamics of Ground State Bimolecular Reactions.
C. Wittig and A. H. Zewail
Chemical Reactions in Clusters, ed. E. R. Bernstein,
Oxford University Press, New York, 1996, p. 64

284. Coherence: A Powerful Concept in the Studies of
Structures and Dynamics.
A. H. Zewail
Laser Phys. **5**, 417 (1995)

285. Direct Observation of The Transition State.
J. C. Polanyi and A. H. Zewail
Acc. Chem. Res. **28**, 119 (1995)

286. Kinetic-Energy, Femtosecond Resolved Reaction
Dynamics: Modes of Dissociation (in Iodobenzene) from
Time-Velocity Correlations.
P. Y. Cheng, D. Zhong, and A. H. Zewail
Chem. Phys. Lett. **237**, 399 (1995)

287. Femtochemistry of Organometallics: Dynamics of Metal-
Metal and Metal-Ligand Bond Cleavage in M2(CO)10.
S. K. Kim, S. Pedersen, and A. H. Zewail
Chem. Phys. Lett. **233**, 500 (1995)

288. Femtosecond Dynamics of Reactions: Elementary
Processes of Controlled Solvation.
A. H. Zewail
Ber. Bunsenges. Phys. Chem. **99**, 474 (1995)

289. Classical Theory of Ultrafast Pump-Probe Spectroscopy:
Applications to I_2 Photodissociation in Ar Solution.
R. M. Whitnell, K. R. Wilson, Y. Yan, and A. H. Zewail
J. Mol. Liquids **61**, 153 (1994)

290. Direct Femtosecond Observation of the Transient
Intermediate in the α-Cleavage Reaction of $(CH_3)_2CO$ to
$2CH_3 + CO$: Resolving the Issue of Concertedness.

S. K. Kim, S. Pedersen, and A. H. Zewail
J. Chem. Phys. **103**, 477 (1995)

291. Solvation Ultrafast Dynamics of Reactions. IX.
Femtosecond Studies of Dissociation and Recombination
of Iodine in Argon Clusters.
J.-K. Wang, Q. Liu, and A. H. Zewail
J. Phys. Chem. **99**, 11309 (1995)

292. Solvation Ultrafast Dynamics of Reactions. X. Molecular
Dynamics Studies of Dissociation, Recombination and
Coherence.
Q. Liu, J.-K. Wang, and A. H. Zewail
J. Phys. Chem. **99**, 11321 (1995)

293. Transition States of Charge-Transfer Reactions:
Femtosecond Dynamics and the Concept of Harpooning in
the Bimolecular Reaction of Benzene with Iodine.
P. Y. Cheng, D. Zhong, and A. H. Zewail
J. Chem. Phys. **103**, 5153 (1995)

294. Microscopic Friction and Solvation in Barrier Crossing:
Isomerization of Stilbene in Size-Selected Hexane
Clusters.
A. A. Heikal, S. H. Chong, J. S. Baskin, and A. H. Zewail
Chem. Phys. Lett. **242**, 380 (1995)

295. Microscopic Solvation and Femtochemistry of Charge-
Transfer Reactions: The Problem of Benzene(s)-Iodine
Binary Complexes and Their Solvent Structures.
P. Y. Cheng, D. Zhong, and A. H. Zewail
Chem. Phys. Lett. **242**, 369 (1995)

296. Femtosecond, Velocity-Gating of Complex Structures in
Solvent Cages.
P. Y. Cheng, D. Zhong, and A. H. Zewail
J. Phys. Chem. **99**, 15733 (1995)

297. Femtochemistry and Max Bodenstein's Impact.
A. H. Zewail
*Springer Series in Chemical Physics, Vol. 61: Gas Phase
Chemical Reaction Systems — Experiments and Models
100 Years After Max Bodenstein*, eds. J. Wolfrum, H.-R.
Volpp, R. Rannacher, and J. Warnatz,Springer-Verlag,
Berlin-Heidelberg, 1996, p. 3

298. Femtosecond Molecular Dynamics of Tautomerization in
Model Base Pairs.
A. Douhal, S. K. Kim, and A. H. Zewail
Nature **378**, 260 (1995)

299. Femtosecond Real-Time Probing of Reactions. XVIII.
Experimental and Theoretical Mapping of Trajectories and
Potentials in the NaI Dissociation Reaction.
P. Cong, G. Roberts, J. L. Herek, A. Mokhtari, and A. H.
Zewail
J. Phys. Chem. **100**, 7832 (1996)

300. Femtochemistry: Advances Over a Decade.
A. H. Zewail
*Femtochemistry: Ultrafast Chemical and Physical
Processes in Molecular Systems*, ed. M. Chergui, World
Scientific, Singapore, 1996, p. 3

301. Femtosecond Control of an Elementary Unimolecular
Reaction from the Transition-State Region.
A. Materny, P. Cong, J. L. Herek, and A. H. Zewail
*Femtochemistry: Ultrafast Chemical and Physical
Processes in Molecular Systems*, ed. M. Chergui, World
Scientific, Singapore, 1996, p. 356

302. Femtosecond Elementary Dynamics of Transition States
and Asymmetric α-Cleavage in Norrish Reactions.
S. K. Kim and A. H. Zewail
Chem. Phys. Lett. **250**, 279 (1996)

303. Proton-Transfer Reaction Dynamics.
A. Douhal, F. Lahmani, and A. H. Zewail
Chem. Phys. **207**, 477 (1996)

304. Femtosecond Real-Time Probing of Reactions. XIX.
Nonlinear (DFWM) Techniques for Probing Transition
States of Uni- and Bimolecular Reactions.
M. Motzkus, S. Pedersen, and A. H. Zewail
J. Phys. Chem. **100**, 5620 (1996)

305. Femtosecond Chemically Activated Reactions: Concept of
Nonstatistical Activation at High Thermal Energies.
S. K. Kim, J. Guo, J. S. Baskin, and A. H. Zewail
J. Phys. Chem. **100**, 9202 (1996)

306. Femtochemistry: Recent Progress in Studies of Dynamics
and Control of Reactions and Their Transition States.
A. H. Zewail
J. Phys. Chem. **100**, 12701 (1996)

307. Femtochemistry: Chemical Reaction Dynamics and Their
Control.
A. H. Zewail
*Advances in Chemical Physics, Vol. 101: Chemical
Reactions and Their Control on the Femtosecond Time
Scale*, eds. P. Gaspard and I. Burghardt,
John Wiley & Sons, New York, 1997, p. 3

308. Femtosecond Real-Time Probing of Reactions. XX.
Dynamics of Twisting, Alignment, and IVR in the *trans*-
Stilbene Isomerization Reaction.
J. S. Baskin, L. Bañares, S. Pedersen, and A. H. Zewail
J. Phys. Chem. **100**, 11920 (1996)

309. Femtochemistry of ICN in Liquids: Dynamics of
Dissociation, Recombination and Abstraction.
C. Wan, M. Gupta, and A. H. Zewail
Chem. Phys. Lett. **256**, 279 (1996)

310. Femtosecond Real-Time Probing of Reactions. XXI. Direct Observation of Transition-State Dynamics and Structure in Charge-Transfer Reactions.
P. Y. Cheng, D. Zhong, and A. H. Zewail
J. Chem. Phys. **105**, 6216 (1996)

311. Femtosecond Real-Time Probing of Reactions. XXII. Kinetic Description of Probe Absorption, Fluorescence, Depletion and Mass Spectrometry.
S. Pedersen and A. H. Zewail
Mol. Phys. **89**, 1455 (1996)

312. Retro-Diels-Alder Femtosecond Reaction Dynamics.
B. A. Horn, J. L. Herek, and A. H. Zewail
J. Am. Chem. Soc. **118**, 8755 (1996)

313. Femtosecond Reaction Dynamics in the Gas-to-Liquid Transition Region: Observation of a Three-Phase Density Dependence.
Q. Liu, C. Wan, and A. H. Zewail
J. Chem. Phys. **105**, 5294 (1996)

314. Structural Effect on the Isomerization Dynamics of *trans*-Stilbenes: IVR, Microcanonical Reaction Rates, and the Nature of the Transition State.
A. A. Heikal, J. S. Baskin, L. Bañares, and A. H. Zewail
J. Phys. Chem. A **101**, 572 (1997)

315. Bimolecular Reactions Observed by Femtosecond Detachment to Aligned Transition States: Inelastic and Reactive Dynamics.
D. Zhong, P. Y. Cheng, and A. H. Zewail
J. Chem. Phys. **105**, 7864 (1996)

316. Solvation Ultrafast Dynamics of Reactions. XI. Dissociation and Caging Dynamics in the Gas-to-Liquid Transition Region.
C. Lienau and A. H. Zewail
J. Phys. Chem. **100**, 18629 (1996)

317. Solvation Ultrafast Dynamics of Reactions. XII. Probing along the Reaction Coordinate and Dynamics in Supercritical Argon.
A. Materny, C. Lienau, and A. H. Zewail
J. Phys. Chem. **100**, 18650 (1996)

318. Solvation Ultrafast Dynamics of Reactions. XIII. Theoretical and Experimental Studies of Wave Packet Reaction Coherence and Its Density Dependence.
Q. Liu, C. Wan, and A. H. Zewail
J. Phys. Chem. **100**, 18666 (1996)

319. Femtosecond Nucleophilic Substitution Reaction Dynamics.
D. Zhong, S. Ahmad, P. Y. Cheng, and A. H. Zewail
J. Am. Chem. Soc. **119**, 2305 (1997)

320. Caging Phenomena in Reactions: Femtosecond Observation of Coherent, Collisional Confinement.
C. Wan, M. Gupta, J. S. Baskin, Z. H. Kim, and A. H. Zewail
J. Chem. Phys. **106**, 4353 (1997)

321. David: An Epitome of Civilized Humanity and Great Scholarship.
A. H. Zewail
Optical, Electrical and Magnetic Properties of Molecules: A Review of the Work of A. D. Buckingham, eds. D. C. Clary and B. J. Orr, Elsevier, Amsterdam, 1997, p. xvii

322. Femtochemistry: Dynamics with Atomic Resolution.
A. H. Zewail
Femtochemistry and Femtobiology: Ultrafast Reaction Dynamics at Atomic-Scale Resolution, ed. V. Sundström, World Scientific, Singapore, 1997, p. 1

323. Clocking Transient Chemical Changes by Ultrafast Electron Diffraction.
J. C. Williamson, J. Cao, H. Ihee, H. Frey, and A. H. Zewail
Nature **386**, 159 (1997)

324. Femtochemistry: Reaction Dynamics with Atomic Resolution.
A. H. Zewail, R. S. Berry, S. A. Rice, and J. Ross
Physical Chemistry, 2 nd ed., Oxford University Press, New York,
2000, p. 896

325. Femtosecond Elimination Reaction Dynamics.
D. Zhong, S. Ahmad, and A. H. Zewail
J. Am. Chem. Soc. **119**, 5978 (1997)

326. Molecular Clusters: Real-Time Dynamics and Reactivity.
J. A. Syage and A. H. Zewail
*Advances in Molecular Vibrations and Collision
Dynamics, Vol. 3:
Molecular Clusters*, eds. J. M. Bowman and Z. Bačić, JAI
Press, Stanford, 1998, p. 1

327. What is Chemistry? 100 Years After J. J. Thomson's
Discovery.
A. H. Zewail
Cambridge Rev. **118**, 65 (1997)

328. Femtosecond Dynamics of Microscopic Friction: Nature of
Coherent versus Diffusive Motion from Gas to Liquid
Density.
J. S. Baskin, M. Gupta, M. Chachisvilis, and A. H. Zewail
Chem. Phys. Lett. **275**, 437 (1997)

329. Ultrafast Electron Diffraction: Structures in Dissociation
Dynamics of $Fe(CO)_5$
H. Ihee, J. Cao, and A. H. Zewail
Chem. Phys. Lett. **281**, 10 (1997)

330. Femtosecond Activation of Reactions and the Concept of
Nonergodic Molecules.
E. W.-G. Diau, J. L. Herek, Z. H. Kim, and A. H. Zewail
Science **279**, 847 (1998)

331. Femtochemistry: Atomic-Scale Resolution of Physical,
     Chemical and Biological Dynamics.
     A. H. Zewail
     Proc. 41st Robert A. Welch Conf. on Chemical Research:
     The Transactinide Elements, Welch Foundation, Houston,
     1997, p. 323

332. Femtosecond Dynamics of Solvation: Microscopic Friction
     and Coherent Motion in Dense Fluids.
     J. S. Baskin, M. Chachisvilis, M. Gupta, and A. H. Zewail
     J. Phys. Chem. A **102**, 4158 (1998)

333. Femtosecond Dynamics of a Hydrogen-Bonded Model
     Base Pair in the Condensed Phase: Double Proton
     Transfer in 7-Azaindole.
     M. Chachisvilis, T. Fiebig, A. Douhal, and A. H. Zewail
     J. Phys. Chem. A **102**, 669 (1998)

334. Femtochemistry in Nanocavities: Reactions in Cyclodextrins.
     A. Douhal, T. Fiebig, M. Chachisvilis, and A. H. Zewail
     J. Phys. Chem. A **102**, 1657 (1998)

335. Femtosecond Real-Time Probing of Reactions. XXIII.
     Studies of Temporal, Velocity, Angular and State
     Dynamics from Transition States to Final Products
     by Femtosecond-Resolved Mass Spectrometry.
     D. Zhong and A. H. Zewail
     J. Phys. Chem. A **102**, 4031 (1998)

336. Solvation Ultrafast Dynamics of Reactions. XIV.
     Molecular Dynamics and *ab Initio* Studies of Charge-
     Transfer Reactions of Iodine in Benzene Clusters.
     J. T. Su and A. H. Zewail
     J. Phys. Chem. A **102**, 4082 (1998)

337. Femtosecond Dynamics of Transition States and the
     Concept of Concertedness: Nitrogen Extrusion of
     Azomethane Reactions.

E. W.-G. Diau, O. K. Abou-Zied, A. A. Scala, and A. H. Zewail
J. Am. Chem. Soc. **120**, 3245 (1998)

338. Freezing Time — in a Femtosecond.
J. S. Baskin and A. H. Zewail
Sci. Spectra **14**, 62 (1998)

339. Femtosecond ß-Cleavage Dynamics: Observation of the Diradical Intermediate in the Nonconcerted Reactions of Cyclic Ethers.
A. A. Scala, E. W.-G. Diau, Z. H. Kim, and A. H. Zewail
J. Chem. Phys. **108**, 7933 (1998)

340. Femtosecond Dynamics of Metalloporphyrins: Electron Transfer and Energy Redistribution.
H.-Z. Yu, J. S. Baskin, B. Steiger, C. Wan, F. C. Anson, and A. H. Zewail
Chem. Phys. Lett. **293**, 1 (1998)

341. Femtochemistry in Nanocavities: Dissociation, Recombination and Vibrational Cooling of Iodine in Cyclodextrin.
M. Chachisvilis, I. Garcia-Ochoa, A. Douhal, and A. H. Zewail
Chem. Phys. Lett. **293**, 153 (1998)

342. Ultrafast Electron Diffraction: Determination of Radical Structure with Picosecond Time Resolution.
J. Cao, H. Ihee, and A. H. Zewail
Chem. Phys. Lett. **290**, 1 (1998)

343. Femtosecond Dynamics of Transition States: The Classical Saddle-Point Barrier Reactions.
K. B. Møller and A. H. Zewail
Chem. Phys. Lett. **295**, 1 (1998); *ibid.* **296**, 619 (1998)

344. Femtosecond Dynamics of Valence-Bond Isomers of Azines: Transition States and Conical Intersections.
D. Zhong, E. W.-G. Diau, T. M. Bernhardt, S. De Feyter,

J. D. Roberts, and A. H. Zewail
Chem. Phys. Lett. **298**, 129 (1998)

345. Ground- and Excited-State Reactions of Norbornene and Isomers: A CASSCF Study and Comparison with Femtosecond Experiments.
S. Wilsey, K. N. Houk, and A. H. Zewail
J. Am. Chem. Soc. **121**, 5772 (1999)

346. Femtosecond Dynamics and Electrocatalysis of the Reduction of O_2: Tetraruthenated Cobalt Porphyrins.
H.-Z. Yu, J. S. Baskin, B. Steiger, F. C. Anson, and A. H. Zewail
J. Am. Chem. Soc. **121**, 484 (1999)

347. Ultrafast Electron Diffraction and Direct Observation of Transient Structures in a Chemical Reaction.
J. Cao, H. Ihee, and A. H. Zewail
Proc. Natl. Acad. Sci. U.S.A. **96**, 338 (1999)

348. Femtosecond Dynamics of the DNA Intercalator Ethidium and Electron Transfer with Mononucleotides in Water.
T. Fiebig, C. Wan, S. O. Kelley, J. K. Barton, and A. H. Zewail
Proc. Natl. Acad. Sci. U.S.A. **96**, 1187 (1999)

349. Femtosecond Dynamics of Dative Bonding: Concepts of Reversible and Dissociative Electron Transfer Reactions.
D. Zhong and A. H. Zewail
Proc. Natl. Acad. Sci. U.S.A. **96**, 2602 (1999)

350. Direct Observation of the Femtosecond Nonradiative Dynamics of Azulene in a Molecular Beam: The Anomalous Behavior in the Isolated Molecule.
E. W.-G. Diau, S. De Feyter, and A. H. Zewail
J. Chem. Phys. **110**, 9785 (1999)

351. Femtosecond Dynamics of Diradicals: Transition States, Entropic Configurations and Stereochemistry.
S. De Feyter, E. W.-G. Diau, A. A. Scala, and A. H. Zewail
Chem. Phys. Lett. **303**, 249 (1999)

352. Conformations and Barriers of Haloethyl Radicals (CH$_2$XCH$_2$, X=F, Cl, Br, I): *Ab Initio* Studies.
H. Ihee, A. H. Zewail, and W. A. Goddard III
J. Phys. Chem. A **103**, 6638 (1999)

353. Femtosecond Dynamics of DNA-Mediated Electron Transfer.
C. Wan, T. Fiebig, S. O. Kelley, C. R. Treadway, J. K. Barton, and A. H. Zewail
Proc. Natl. Acad. Sci. U.S.A. **96**, 6014 (1999)

354. Femtosecond Dynamics of Retro Diels-Alder Reactions: The Concept of Concertedness.
E. W.-G. Diau, S. De Feyter, and A. H. Zewail
Chem. Phys. Lett. **304**, 134 (1999)

355. Femtosecond Dynamics of Pyridine in the Condensed Phase: Valence Isomerization by Conical Intersections.
M. Chachisvilis and A. H. Zewail
J. Phys. Chem. A **103**, 7408 (1999)

356. Femtosecond Dynamics of Double Proton Transfer in a Model DNA Base Pair: 7-Azaindole Dimers in the Condensed Phase.
T. Fiebig, M. Chachisvilis, M. Manger, A. H. Zewail, A. Douhal, I. Garcia-Ochoa, and A. de La Hoz Ayuso
J. Phys. Chem. A **103**, 7419 (1999)

357. Femtosecond Real-Time Probing of Reactions. XXIV. Time, Velocity and Orientation Mapping of the Dynamics of Dative Bonding in Bimolecular Electron Transfer Reactions.

D. Zhong, T. M. Bernhardt, and A. H. Zewail
J. Phys. Chem. A **103**, 10093 (1999)

358. The Concept of Coherent Resonances in the Nuclear
    Motion of Bimolecular Collisions: Femtosecond
    Probing and the Classical Picture.
    K. B. Møller and A. H. Zewail
    Chem. Phys. Lett. **309**, 1 (1999)

359. Femtosecond Dynamics of Ground-State Vibrational
    Motion and Energy Flow: Polymers of Diacetylene.
    (A) Vierheilig, T. Chen, P. Waltner, W. Kiefer, A.
    (B) Materny, and A. H. Zewail
    Chem. Phys. Lett. **312**, 349 (1999)

360. Femtosecond Dynamics of Norrish Type-II Reactions:
    Nonconcerted Hydrogen-Transfer and Diradical
    Intermediacy.
    S. De Feyter, E. W.-G. Diau, and A. H. Zewail
    Angew. Chem. Int. Ed. Engl. **39**, 260 (2000)

361. Femtosecond Dynamics of Dioxygen-Picket-Fence
    Cobalt Porphyrins: Ultrafast Release of O_2 and the
    Nature of Dative Bonding.
    B. Steiger, J. S. Baskin, F. C. Anson, and A. H. Zewail
    Angew. Chem. Int. Ed. Engl. **39**, 257 (2000)

362. Femtosecond Dynamics of Hydrogen Elimination:
    Benzene Formation from Cyclohexadiene.
    S. De Feyter, E. W.-G. Diau, and A. H. Zewail
    Phys. Chem. Chem. Phys. **2**, 877 (2000)

363. Femtosecond Observation of Benzyne Intermediates in a
    Molecular Beam: Bergman Rearrangement in the
    Isolated Molecule.
    E. W.-G. Diau, J. Casanova, J. D. Roberts, and A. H. Zewail
    Proc. Natl. Acad. Sci. U.S.A. **97**, 1376 (2000)

364. Femtosecond Linear Dichroism of DNA-Intercalating
Chromophores: Solvation and Charge Separation
Dynamics of [Ru(phen)$_2$dppz]$_{2+}$ Systems.
B. Önfelt, P. Lincoln, B. Nordén, J. S. Baskin, and A. H.
Zewail
Proc. Natl. Acad. Sci. U.S.A. **97**, 5708 (2000)

365. Femtochemistry: Atomic-Scale Dynamics of the
Chemical Bond Using Ultrafast Lasers.
A. H. Zewail
*Les Prix Nobel, The Nobel Prizes 1999: Nobel Prizes,
Presentations, Biographies, and Lectures*, ed. T.
Frängsmyr, Almquist & Wiksell, Stockholm, 2000, p. 103;
Nobel Lectures in Chemistry 1996–2000, ed. I. Grenthe,
World Scientific, Singapore, 2003, p. 262

Fig. 11.3   Ahmed H. Zewail, left, receiving the Nobel Prize in chemistry from
Swedish King Carl XVI Gustaf, right.

366. Femtochemistry: Atomic-Scale Dynamics of the
     Chemical Bond Using Ultrafast Lasers.
     A. H. Zewail
     Angew. Chem. Intl. Ed. Engl. **39**, 2586 (2000);
     Angew. Chem. (German) **112**, 2688 (2000);
     Femtochemistry, eds. F. C. De Schryver, S. De Feyter,
     and G. Schweitzer, Wiley-VCH, Weinheim, 2001, p. 1

367. Femtochemistry: Atomic-Scale Dynamics of the Chemical Bond.
     A. H. Zewail
     J. Phys. Chem. A **104**, 5660 (2000)

368. Femtosecond Direct Observation of Charge Transfer
     between Bases in DNA.
     C. Wan, T. Fiebig, O. Schiemann, J. K. Barton, and
     A. H. Zewail
     Proc. Natl. Acad. Sci. U.S.A. **97**, 14052 (2000)

369. On the Role of Coherence in the Transition from
     Kinetics to Dynamics: Theory and Application to
     Femtosecond Unimolecular Reactions.
     K. B. Møller, N. E. Henriksen, and A. H. Zewail
     J. Chem. Phys. **113**, 10477 (2000)

370. Femtosecond Activation of Reactions: The Concepts of
     Nonergodic Behavior and Reduced-Space Dynamics.
     K. B. Møller and A. H. Zewail
     *Essays in Contemporary Chemistry: From Molecular
     Structure to towards Biology*, eds. G. Quinkert and M.
     V. Kisakürek, VHCA, Wiley-VCH, Zürich-Weinheim,
     2001, p. 157

371. Organic Femtochemistry: Diradicals, Theory and
     Experiments.
     S. De Feyter, E. W.-G. Diau, and A. H. Zewail
     Femtochemistry, eds. F. C. De Schryver, S. De Feyter
     and G. Schweitzer, Wiley-VCH, Weinheim, 2001, p. 97

372. Femtochemistry: Past, Present, and Future.
   A. H. Zewail
   Pure Appl. Chem. **72**, 2219 (2000)

373. Femtosecond Atomic-Scale Stroboscopy: From Physics
   to Chemistry and on to Biology.
   A. H. Zewail
   *Proc. SIF, Vol. 71: Atoms, Molecules, and Quantum
   Dots in Laser Fields — Fundamental Processes*, eds. N.
   Bloembergen, N. Rahman, and A. Rizzo, SIF, Bologna,
   2001, p. 1

374. Direct Imaging of Transient Molecular Structures with
   Ultrafast Diffraction.
   H. Ihee, V. A. Lobastov, U. M. Gomez, B. M. Goodson,
   R. Srinivasan, C.-Y. Ruan, and A. H. Zewail
   Science **291**, 458 (2001)

375. Femtosecond Studies of Protein-Ligand Hydrophobic
   Binding and Dynamics: Human Serum Albumin.
   D. Zhong, A. Douhal, and A. H. Zewail
   Proc. Natl. Acad. Sci. U.S.A. **97**, 14056 (2000)

376. Direct Observation of Resonance Motion in Complex
   Elimination Reactions: Femtosecond Coherent
   Dynamics in Reduced Space.
   C. Kötting, E. W.-G. Diau, J. E. Baldwin, and
   A. H. D. Zewail
   J. Phys. Chem. A **105**, 1677 (2001)

F.1 *Dr. Zewail's publications on his brilliant research works in Caltech on femtochemistry and femtobiology*

377. Femtosecond Studies of Protein-DNA Binding and
   Dynamics: Histone I.
   D. Zhong, S. K. Pal, and A. H. Zewail
   Chem. Phys. Chem. **2**, 219 (2001)

378. Molecular Structure and Orientation: Concepts from Femtosecond Dynamics.
J. S. Baskin and A. H. Zewail
J. Phys. Chem. A **105**, 3680 (2001)

379. Ultrafast Electron Diffraction of Transient $Fe(CO)_4$: Determination of Molecular Structure and Reaction Pathway.
H. Ihee, J. Cao, and A. H. Zewail
Angew. Chem. Int. Ed. Engl. **40**, 1532 (2001)

380. Femtochemistry of Norrish Type-I Reactions. I. Experimental and Theoretical Studies of Acetone and Related Ketones on the S_1 Surface.
E. W.-G. Diau, C. Kötting, and A. H. Zewail
Chem. Phys. Chem. **2**, 273 (2001)

381. Femtochemistry of Norrish Type-I Reactions. II. The Anomalous Predissociation Dynamics of Cyclobutanone on the S_1 Surface.
E. W.-G. Diau, C. Kötting, and A. H. Zewail
Chem. Phys. Chem. **2**, 294 (2001)

382. Femtochemistry: Recent Progress in Studies of Dynamics and Control of Reactions and Their Transition States.
A. H. Zewail
Atomic and Molecular Beams: The State of the Art 2000, ed. R. Campargue, Springer-Verlag, Berlin-Heidelberg, 2001, p. 415

383. Femtosecond Transition State Dynamics of *cis*-Stilbene.
T. Baumert, T. Frohnmeyer, B. Kiefer, P. Niklaus, M. Strehle, G. Gerber, and A. H. Zewail
Appl. Phys. B **72**, 105 (2001)

384. CF_2XCF_2X and CF_2XCF_2 • Radicals (X = Cl, Br, I): *Ab Initio* and DFT Studies and Comparisons with Experiments.

H. Ihee, J. Kua, W. A. Goddard III, and A. H. Zewail
J. Phys. Chem. A **105**, 3623 (2001)

385. Time and Matter: Science at New Limits.
A. H. Zewail
Pontific. Acad. Sci. Scripta Varia **99**, 426 (2001)

386. The Uncertainty Paradox: The Fog That Was Not.
A. H. Zewail
Nature **412**, 279 (2001)

387. Freezing Atoms in Motion: Principles of Femtochemistry and Demonstration by Laser Stroboscopy.
J. S. Baskin and A. H. Zewail
J. Chem. Educ. **78**, 737 (2001); *The Life and Scientific Legacy of George Porter*, eds. D. Phillips and J. Barber, Imperial College Press, London, 2006, p. 340

388. Femtochemistry of Mass-Selected Negative-Ion Clusters of Dioxygen: Charge-Transfer and Solvation Dynamics.
D. H. Paik, T. M. Bernhardt, N. J. Kim, and A. H. Zewail
J. Chem. Phys. **115**, 612 (2001)

389. Ultrafast Diffraction and Structural Dynamics: The Nature of Complex Molecules Far from Equilibrium.
C.-Y. Ruan, V. A. Lobastov, R. Srinivasan, B. M. Goodson, H. Ihee, and A. H. Zewail
Proc. Natl. Acad. Sci. U.S.A. **98**, 7117 (2001)

390. The Anticancer Drug-DNA Complex: Femtosecond Primary Dynamics for Anthracycline Antibiotics Function.
X. Qu, C. Wan, H.-C. Becker, D. Zhong, and A. H. Zewail
Proc. Natl. Acad. Sci. U.S.A. **98**, 14212 (2001)

391. Femtochemistry of Norrish Type-I Reactions. III. Highly-Excited Ketones, Theoretical.

E. W.-G. Diau, C. Kötting, T. I. Sølling, and A. H. Zewail
Chem. Phys. Chem. **3**, 57 (2002)

392. Femtochemistry of Norrish Type-I Reactions. IV. Highly-Excited Ketones, Experimental.
T. I. Sølling, E. W.-G. Diau, C. Kötting, S. De Feyter, and A. H. Zewail
Chem. Phys. Chem. **3**, 79 (2002)

393. Ultrafast Diffraction of Transient Molecular Structures in Radiationless Transitions.
V. A. Lobostov, R. Srinivasan, B. M. Goodson, C.-Y. Ruan, J. S. Feenstra, and A. H. Zewail
J. Phys. Chem. A **105**, 11159 (2001)

394. Chemistry at the Uncertainty Limit.
A. H. Zewail
Angew. Chem. Int. Ed. Engl. **40**, 4371 (2001);
Angew. Chem. (German) **113**, 4501 (2001)

395. DNA/RNA Nucleotides and Nucleosides: Direct Measurement of Excited-State Lifetimes by Femtosecond Fluorescence Up-Conversion.
J. Peon and A. H. Zewail
Chem. Phys. Lett. **348**, 255 (2001)

396. Femtosecond Dynamics of Flavoproteins: Charge Separation and Recombination in Riboflavine (Vitamin B_2)-Binding Protein and in Glucose Oxidase Enzyme.
D. Zhong and A. H. Zewail
Proc. Natl. Acad. Sci. U.S.A. **98**, 11867 (2001)

397. Femtosecond Dynamics of Drug-Protein Complex: Daunomycin with Apo Riboflavin-Binding Protein.
D. Zhong, S. K. Pal, C. Wan, and A. H. Zewail
Proc. Natl. Acad. Sci. U.S.A. **98**, 11873 (2001)

398. Coherent Dynamics in Complex Elimination Reactions: Experimental and Theoretical Femtochemistry of 1,3-Dibromopropane and Related Systems.
C. Kötting, E. W.-G. Diau, T. I. Sølling, and A. H. Zewail
J. Phys. Chem. A **106**, 7530 (2002)

399. Kinetics Modeling of Dynamics: The Case of Femtosecond-Activated Direct Reactions.
K. B. Møller and A. H. Zewail
Chem. Phys. Lett. **351**, 281 (2002)

400. Femtosecond Dynamics of Rubredoxin: Tryptophan Solvation and Resonance Energy Transfer in the Protein.
D. Zhong, S. K. Pal, D. Zhang, S. I. Chan, and A. H. Zewail
Proc. Natl. Acad. Sci. U.S.A. **99**, 13 (2002)

401. Ultrafast Electron Diffraction and Structural Dynamics: Transient Intermediates in the Elimination Reaction of C2F4I2.
H. Ihee, B. M. Goodson, R. Srinivasan, V. A. Lobastov, and A. H. Zewail
J. Phys. Chem. A **106**, 4087 (2002)

402. Ultrafast Electron Diffraction of Transient Cyclopentadienyl Radical: A Dynamic Pseudorotary Structure.
H. Ihee, J. S. Feenstra, J. Cao, and A. H. Zewail
Chem. Phys. Lett. **353**, 325 (2002)

403. Orientation Dynamics and Molecular Structures from Gas Phase to Condensed Media.
J. S. Baskin and A. H. Zewail
Femtochemistry and Femtobiology: Ultrafast Dynamics In Molecular Science, eds. A. Douhal and J. Santamaria, World Scientific, Singapore, 2002, p. 3

404. Biological Water at the Protein Surface: Dynamical
Solvation Probed Directly with Femtosecond Resolution.
S. K. Pal, J. Peon, and A. H. Zewail
Proc. Natl. Acad. Sci. U.S.A. **99**, 1763 (2002)

405. Ultrafast Dynamics of Porphyrins in the Condensed
Phase. I. Free Base Tetraphenylporphyrin.
J. S. Baskin, H.-Z. Yu, and A. H. Zewail
J. Phys. Chem. A **106**, 9837 (2002)

406. Ultrafast Dynamics of Porphyrins in the Condensed
Phase. II. Zinc Tetraphenylporphyrin.
H.-Z. Yu, J. S. Baskin, and A. H. Zewail
J. Phys. Chem. A **106**, 9845 (2002)

407. Femtosecond Studies of Protein-Ligand Hydrophobic
Binding and Dynamics: Human Serum Albumin.
A. Douhal, D. Zhong, and A. H. Zewail
*Femtochemistry and Femtobiology: Ultrafast Dynamics
in Molecular Science*, eds. A. Douhal and J. Santamaria,
World Scientific, Singapore, 2002, p. 731

408. Linus Pauling: A Great Pharaoh of Chemistry.
A. H. Zewail
*World Scientific Series in XXth Century Chemistry, Vol.
10: Linus Pauling — Selected Scientific Papers*, eds. B.
Kamb, L. Pauling Kamb, P. J. Pauling, A. Kamb, and
L. Pauling Jr., World Scientific, Singapore, 2001, p. vii

409. Femtochemistry and Femtobiology.
A. H. Zewail
*Femtochemistry and Femtobiology: Ultrafast Dynamics
in Molecular Science*, eds. A. Douhal and J. Santamaria,
World Scientific, Singapore, 2002, p. v

410. Femtosecond Charge Transfer Dynamics of a Modified
DNA Base: 2-Aminopurine in Complexes with
Nucleotides.

T. Fiebig, C. Wan, and A. H. Zewail
Chem. Phys. Chem. **3**, 781 (2002)

411. Biological Water: Femtosecond Dynamics of
Macromolecular Hydration.
S. K. Pal, J. Peon, B. Bagchi, and A. H. Zewail
J. Phys. Chem. B **106**, 12376 (2002)

412. Molecular Recognition of Oxygen by Protein Mimics:
Dynamics on the Femtosecond to Microsecond Time Scale.
S. Zou, J. S. Baskin, and A. H. Zewail
Proc. Natl. Acad. Sci. U.S.A. **99**, 9625 (2002)

413. Ultrafast Decay and Hydration Dynamics of DNA Bases
and Mimics.
S. K. Pal, J. Peon, and A. H. Zewail
Chem. Phys. Lett. **363**, 57 (2002)

414. Hydration at the Surface of the Protein Monellin:
Dynamics with Femtosecond Resolution.
J. Peon, S. K. Pal, and A. H. Zewail
Proc. Natl. Acad. Sci. U.S.A. **99**, 10964 (2002)

415. Ultrafast Surface Hydration Dynamics and Expression
of Protein Functionality: α-Chymotrypsin.
S. K. Pal, J. Peon, and A. H. Zewail
Proc. Natl. Acad. Sci. U.S.A. **99**, 15297 (2002)

416. Femtosecond Dynamics of Solvated Oxygen Anions. I.
Bifurcated Electron Transfer Dynamics Probed by
Photoelectron Spectroscopy.
D. H. Paik, N. J. Kim, and A. H. Zewail
J. Chem. Phys. **118**, 6923 (2003)

417. Femtosecond Dynamics of Solvated Oxygen Anions. II.
Nature of Dissociation and Caging in Finite-Sized
Clusters.
N. J. Kim, D. H. Paik, and A. H. Zewail
J. Chem. Phys. **118**, 6930 (2003)

418. Femtochemistry of *trans*-Azomethane: A Combined
     Experimental and Theoretical Study.
     E. W.-G. Diau and A. H. Zewail
     Chem. Phys. Chem. 4, 445 (2003)

419. Ultrafast Electron Diffraction: Complex Landscapes of
     Molecular Structures in Thermal and Light-Mediated
     Reactions.
     B. M. Goodson, C.-Y. Ruan, V. A. Lobastov, R.
     Srinivasan, and A. H. Zewail
     Chem. Phys. Lett. 374, 417 (2003)

420. The RNA-Protein Complex: Direct Probing of the
     Interfacial Recognition Dynamics and Its Correlation
     with Biological Functions.
     T. Xia, H.-C. Becker, C. Wan, A. Frankel, R. W.
     Roberts, and A. H. Zewail
     Proc. Natl. Acad. Sci. U.S.A. 100, 8119 (2003)

421. Water at DNA Surfaces: Ultrafast Dynamics in Minor
     The RNA-Protein Complex: Direct Probing of the
     Groove Recognition.
     S. K. Pal, L. Zhao, and A. H. Zewail
     Proc. Natl. Acad. Sci. U.S.A. 100, 8113 (2003)

422. Ultrafast Electron Diffraction (UED): A New
     Development for the 4D Determination of Transient
     Molecular Structures.
     R. Srinivasan, V. A. Lobastov, C.-Y. Ruan, and A. H.
     Zewail
     Helv. Chim. Acta 86, 1763 (2003)

423. Dynamics of Molecules Near Ionization.
     T. I. Sølling, C. Kötting, and A. H. Zewail
     J. Phys. Chem. A 107, 10872 (2003)

424. Ultrafast Electron Crystallography: Transient Structures
of Molecules, Surfaces and Phase Transitions.
C.-Y. Ruan, F. Vigliotti, V. A. Lobastov, S. Chen, and
A. H. Zewail
Proc. Natl. Acad. Sci. U.S.A. **101**, 1123 (2004)

425. Dynamics of Water Near a Protein Surface.
S. M. Bhattacharyya, Z.-G. Wang, and A. H. Zewail
J. Phys. Chem. B **107**, 13218 (2003)

426. Dynamics of Water in Biological Recognition.
S. K. Pal and A. H. Zewail
Chem. Rev. **104**, 2099 (2004);
Introduction: Femtochemistry
M. Dantus, A. H. Zewail, *ibid.* **104**, 1717 (2004)

427. Ultrafast Electron Diffraction: From the Gas Phase to
the Condensed Phase with Picosecond and Femtosecond
Resolution.
V. A. Lobastov, R. Srinivasan, F. Vigliotti, C.-Y. Ruan,
J. S. Feenstra, S. Chen, S. T. Park, S. Xu, and A. H. Zewail
*Springer Series in Optical Sciences, Vol. 95: Ultrafast
Optics IV,* eds. F. Krausz, G. Korn, P. Corkum, I. A.
Walmsley, Springer, New York, 2004, p. 419

428. Site and Sequence Selective Ultrafast Hydration of
DNA.
S. K. Pal, L. Zhao, T. Xia, and A. H. Zewail
Proc. Natl. Acad. Sci. U.S.A. **100**, 13746 (2003)

429. Ultrafast Dynamics in DNA-Mediated Electron
Transfer: Base Gating and the Role of Temperature.
M. A. O'Neill, H.-C. Becker, C. Wan, J. K. Barton, and
A. H. Zewail
Angew. Chem. Int. Ed. Engl. **42**, 5896 (2003)

430. Dynamics of Ordered Water in Interfacial Enzyme
     Recognition: Bovine Pancreatic Phospholipase A2.
     L. Zhao, S. K. Pal, T. Xia, and A.H. Zewail
     Angew. Chem. Int. Ed. Engl. **43**, 59 (2004)

431. Ultrafast Electron Diffraction and Transient Complex
     Structures: From Gas Phase to Crystallography.
     A. H. Zewail
     *Femtochemistry and Femtobiology: Ultrafast Events in
     Molecular Science*, eds. M. M. Martin and J. T. Hynes,
     Elsevier, Amsterdam, 2004, p. 3

432. Ultrafast Electron Crystallography of Interfacial Water.
     C.-Y. Ruan, V. A. Lobastov, F. Vigliotti, S. Chen, and
     A. H. Zewail
     Science **304**, 80 (2004)

433. Direct Determination of Hydrogen-Bonded Structures in
     Resonant and Tautomeric Reactions Using Ultrafast
     Electron Diffraction.
     R. Srinivasan, J. S. Feenstra, S. T. Park, S. Xu, and A. H. Zewail
     J. Am. Chem. Soc. **126**, 2266 (2004)

434. Ultrafast Electron Crystallography of Surface Structural
     Dynamics with Atomic-Scale Resolution.
     F. Vigliotti, S. Chen, C.-Y. Ruan, V. A. Lobastov, and
     A. H. Zewail
     Angew. Chem. Int. Ed. Engl. **43**, 2705 (2004)

435. Enzyme Functionality and Solvation of Subtilisin
     Carlsberg: from Hours to Femtoseconds.
     J. K. A. Kamal, T. Xia, S. K. Pal, L. Zhao, and A. H. Zewail
     Chem. Phys. Lett. **387**, 209 (2004)

436. The Presolvated Electron in Water: Can It Be
     Scavenged at Long Range?
     Q.-B. Lu, J. S. Baskin, and A. H. Zewail
     J. Phys. Chem. B **108**, 10509 (2004)

437. The Transition State of Thermal Organic Reactions:
Direct Observation in Real Time.
D. H. Paik, D.-S. Yang, I-R. Lee, and A. H. Zewail
Angew. Chem. Int. Ed. Engl. **43**, 2830 (2004)

438. Ultrafast Electron Diffraction: Structural Dynamics of
the Elimination Reaction of Acetylacetone.
S. Xu, S. T. Park, J. S. Feenstra, R. Srinivasan, and A. H. Zewail
J. Phys. Chem. A **108**, 6650 (2004)

439. Diffraction, Crystallography, and Microscopy Beyond
Three Dimensions: Structural Dynamics in Space and
Time.
A. H. Zewail
Phil. Trans. R. Soc. A **364**, 315 (2005)

440. Femtosecond Laser Spectroscopy.
A. H. Zewail
Femtosecond Laser Spectroscopy, ed. P. Hannaford,
Springer, New York, 2005, p. xv

441. Ultrafast Hydration Dynamics in Protein Unfolding:
Human Serum Albumin.
J. K. A. Kamal, L. Zhao, and A. H. Zewail
Proc. Natl. Acad. Sci. U.S.A. **101**, 13411 (2004)

442. Electrons in Finite-Sized Water Cavities: Hydration
Dynamics Observed in Real Time.
D. H. Paik, I-R. Lee, D.-S. Yang, J. S. Baskin, and A. H.
Zewail
Science **306**, 672 (2004)

443. Structures and Dynamics of Self-Assembled Surface
Monolayers Observed by Ultrafast Electron
Crystallography.
C.-Y. Ruan, D.-S. Yang, and A. H. Zewail
J. Am. Chem. Soc. **126**, 12797 (2004)

444. Human Myoglobin Recognition of Oxygen: Dynamics of the Energy Landscape.
Y. Wang, J. S. Baskin, T. Xia, and A. H. Zewail
Proc. Natl. Acad. Sci. U.S.A. **101**, 18000 (2004)

F.2 *Dr. Zewail's publication on his brilliant research works in Caltech on 4-D microscopy*

445. Modern Trends in Physics Research.
A. H. Zewail
Proc. AIP, Vol. 748: Modern Trends in Physics Research, ed. L. El-Nadi,
AIP, Melville-New York, 2005, p. ix

446. Dark Structures in Molecular Radiationless Transitions Determined by Ultrafast Diffraction.
R. Srinivasan, J. S. Feenstra, S. T. Park, S. Xu, and A. H. Zewail
Science **307**, 558 (2005)

447. Flash Photolysis and George Porter.
A. H. Zewail
The Life and Scientific Legacy of George Porter, eds. D. Phillips and J. Barber, Imperial College Press, London, 2006, p. 335

448. Norman Davidson.
H. A. Lester and A. H. Zewail
Biogr. Mem. Natl. Acad. Sci. U.S.A. **86**, 61 (2005)

449. Atomic-Scale Dynamical Structures of Fatty Acid Bilayers Observed by Ultrafast Electron Crystallography.
S. Chen, M. T. Seidel, and A. H. Zewail
Proc. Natl. Acad. Sci. U.S.A. **102**, 8854 (2005)

450. Four-Dimensional Ultrafast Electron Microscopy.
V. A. Lobastov, R. Srinivasan, and A. H. Zewail
Proc. Natl. Acad. Sci. U.S.A. **102**, 7069 (2005)

451. Ultrafast Unequilibrated Charge Transfer: A New Channel in the Quenching of Fluorescent Biological Probes.
C. Wan, T. Xia, H.-C. Becker, and A. H. Zewail
Chem. Phys. Lett. **412**, 158 (2005)

452. RNA-Protein Recognition: Single-Residue Ultrafast Dynamical Control of Structural Specificity and Function.
T. Xia, C. Wan, R. W. Roberts, and A. H. Zewail
Proc. Natl. Acad. Sci. U.S.A. **102**, 13013 (2005)

453. Ultrafast Electron Diffraction: Dynamical Structures on Complex Energy Landscapes.
D. Shorokhov, S. T. Park, and A. H. Zewail
Chem. Phys. Chem. **6**, 2228 (2005)

454. Ultrafast Electron Diffraction: Oriented Molecular Structures in Space and Time.
J. S. Baskin and A. H. Zewail
Chem. Phys. Chem. **6**, 2261 (2005)

455. 4D Ultrafast Electron Diffraction, Crystallography, and Microscopy.
A. H. Zewail
Annu. Rev. Phys. Chem. **57**, 65 (2006)

456. Excited State Molecular Structures and Reactions Directly Determined by Ultrafast Electron Diffraction.
J. S. Feenstra, S. T. Park, and A. H. Zewail
J. Chem. Phys. **123**, 221104 (2005)

457. Dynamics of Clusters: From Elementary to Biological Structures.
P.-Y. Cheng, J. S. Baskin, and A. H. Zewail
Proc. Natl. Acad. Sci. U.S.A. **103**, 10570 (2006)

458. Determining Molecular Structures and Conformations
     Directly from Electron Diffraction Using a Genetic
     Algorithm.
     S. Habershon and A. H. Zewail
     Chem. Phys. Chem. 7, 353 (2006)

459. Primary Steps of the Photoactive Yellow Protein:
     Isolated Chromophore Dynamics and Protein Directed
     Function.
     I-R. Lee, W. Lee, and A. H. Zewail
     Proc. Natl. Acad. Sci. U.S.A. **103**, 258 (2006)

460. Ultrafast Solvation Dynamics of Human Serum
     Albumin: Correlations with Conformational Transitions
     and Site-Selected Recognition.
     W. Qiu, L. Zhang, O. Okobiah, Y. Yang, L. Wang, D.
     Zhong, and A. H. Zewail
     J. Phys. Chem. B **110**, 10540 (2006)

461. Helix-to-Coil Transitions in Proteins: Helicity
     Resonance in Ultrafast Electron Diffraction.
     M. M. Lin, D. Shorokhov, and A. H. Zewail
     Chem. Phys. Lett. **420**, 1 (2006)

462. Physical Biology: The Next 50 Years.
     A. H. Zewail
     Biotechnol. Annu. Rev. Vol. 12, ed. M. R. El-Gewely,
     Elsevier, Amsterdam, 2006, p. v

463. 4D Structural Dynamics.
     A. H. Zewail
     *Femtochemistry VII: Fundamental Ultrafast Processes
     in Chemistry, Physics, and Biology*, eds. A. W.
     Castleman Jr. and M. L. Kimble, Elsevier, Amsterdam,
     2006, p. 3

464. Ultrafast Electron Diffraction: Excited State Structures and Chemistries of Aromatic Carbonyls.
S. T. Park, J. S. Feenstra, and A. H. Zewail
J. Chem. Phys. **124**, 174707 (2006)

465. Ultrafast Vectorial and Scalar Dynamics of Ionic Clusters: Azobenzene Solvated by Oxygen.
D. H. Paik, J. S. Baskin, N. J. Kim, and A. H. Zewail
J. Chem. Phys. **125**, 133408 (2006)

466. Ultrafast T-Jump in Water: Studies of Conformation and Reaction Dynamics at the Thermal Limit.
H. Ma, C. Wan, and A. H. Zewail
J. Am. Chem. Soc. **128**, 6338 (2006)

467. Oriented Ensembles in Ultrafast Electron Diffraction.
J. S. Baskin and A. H. Zewail
Chem. Phys. Chem. **7**, 1562 (2006)

468. Ultrafast Electron Diffraction: Structural Dynamics of Molecular Rearrangement in the NO Release from Nitrobenzene.
Y. He, A. Gahlmann, J. S. Feenstra, S. T. Park, and A. H. Zewail
Chem. Asian J. **1–2**, 56 (2006)

469. Ultrafast Photoisomerization of the Photoactive Yellow Protein Chromophore in Solution: Influence of the Protonation State.
A. Espagne, D. H. Paik, P. Changenet-Barret, M. M. Martin, and A. H. Zewail
Chem. Phys. Chem. **7**, 1717 (2006)

470. Nonequilibrium Dynamics and Structure of Interfacial Ice.
O. Andreussi, D. Donadio, M. Parrinello, and A. H. Zewail
Chem. Phys. Lett. **426**, 115 (2006)

471. Ultrafast Electron Crystallography of Phospholipids.
S. Chen, M. T. Seidel, and A. H. Zewail
Angew. Chem., Intl. Ed. Engl. **45**, 5154 (2006)

472. Protein Surface Hydration Mapped by Site-Specific
Mutations.
W. Qiu, Y.-T. Kao, L. Zhang, Y. Yang, L. Wang, W. E.
Stites, D. Zhong, and A. H. Zewail
Proc. Natl. Acad. Sci. U.S.A. **103**, 13979 (2006)

473. Breaking Resolution Limits in Ultrafast Electron
Diffraction and Microscopy.
P. Baum and A. H. Zewail
Proc. Natl. Acad. Sci. U.S.A. **103**, 16105 (2006)

474. The Remarkable Phenomena of Hydrogen Transfer.
A. H. Zewail
Hydrogen Transfer Reactions, eds. J. T. Hynes, J. P.
Klinman, H.-H. Limbach, R. L. Schowen, Wiley-VCH,
Weinheim, 2007, p. v

475. Four-Dimensional Ultrafast Electron Microscopy of
Phase Transitions.
M. S. Grinolds, V. A. Lobastov, J. Weissenrieder, and
A. H. Zewail
Proc. Natl. Acad. Sci. U.S.A. **103**, 18427 (2006)

476. Nonequilibrium Phase Transitions in Cuprates Observed
by Ultrafast Electron Crystallography.
N. Gedik, D.-S. Yang, G. Logvenov, I. Bozovic, and
A. H. Zewail
Science **316**, 425 (2007)

477. Visualizing Complexity: Development of 4D
Microscopy and Diffraction for Imaging in Space and
Time.
A. H. Zewail
Visions of Discovery: New Light on Physics,

Cosmology, and Consciousness, eds. R. Y. Chiao, W. D. Phillips, A. J. Leggett, M. L. Cohen, and C. L. Harper, Jr., Cambridge University Press, London, 2008

478. DNA Folding and Melting Observed in Real Time Redefine the Energy Landscape.
 H. Ma, C. Wan, A. Wu, and A. H. Zewail
 Proc. Natl. Acad. Sci. U.S.A. **104**, 712 (2007)

479. Ultrafast Electron Crystallography. I. Nonequilibrium Dynamics of Nanometer-Scale Structures.
 D.-S. Yang, N. Gedik, and A. H. Zewail
 J. Phys. Chem. C **111**, 4889 (2007)

480. Ultrafast Electron Crystallography. II. Surface Adsorbates of Crystalline Fatty Acids and Phospholipids.
 M. T. Seidel, S. Chen, and A. H. Zewail
 J. Phys. Chem. C **111**, 4920 (2007)

481. Double Proton Transfer Dynamics of Model DNA Base Pairs in the Condensed Phase.
 O.-H. Kwon and A. H. Zewail
 Proc. Natl. Acad. Sci. U.S.A. **104**, 8703 (2007)

482. Ultrafast Electron Crystallography. III. Theoretical Modeling of Structural Dynamics.
 J. Tang, D.-S. Yang, and A. H. Zewail
 J. Phys. Chem. C **111**, 8957 (2007)

483. Voyages with the Master.
 A. H. Zewail
 Turning Points in Solid-State, Materials and Surface Science: A book in Celebration of the Life and Work of Sir John Meurig Thomas, eds. K. D. M. Harris and P. P. Edwards, RSC, Cambridge, 2008, p. 3

484. Dynamics of Electrons in Ammonia Cages: The Discovery System of Solvation.
I-R. Lee, W. Lee, and A. H. Zewail
Chem. Phys. Chem. **9**, 83 (2008)

485. Ultrafast Light-Induced Response of Photoactive Yellow Protein Chromophore Analogues.
A. Espagne, D. H. Paik, P. Changenet-Barret, P. Plaza, M. M. Martin, and A. H. Zewail
Photochem. Photobiol. Sci. **6**, 780 (2007)

486. Atomic-Scale Imaging in Real and Energy Space Developed in Ultrafast Electron Microscopy.
H. S. Park, J. S. Baskin, O.-H. Kwon, and A. H. Zewail
Nano Lett. **7**, 2545 (2007)

487. Ultrafast Electron Microscopy (UEM): 4D Imaging and Diffraction of Nanostructures during Phase Transitions.
V. A. Lobastov, J. Weissenrieder, J. Tang, and A. H. Zewail
Nano Lett. **7**, 2552 (2007)

488. 4D Visualization of Transitional Structures in Phase Transitions by ElectronDiffraction.
P. Baum, D.-S. Yang, and A. H. Zewail
Science **318**, 788 (2007)

489. Picosecond Fluctuating Protein Energy Landscape Mapped by Pressure-Temperature Molecular Dynamics Simulation.
L. Meinhold, J. C. Smith, A. Kitao, and A. H. Zewail
Proc. Natl. Acad. Sci. U.S.A. **104**, 17261 (2007)

490. Controlled Nanoscale Mechanical Phenomena Discovered with UltrafastElectron Microscopy.
D. J. Flannigan, V. A. Lobastov, and A. H. Zewail
Angew. Chem., Int. Ed. Engl. **46**, 9206 (2007)

491. Attosecond Electron Pulses for 4D Diffraction and Microscopy.
P. Baum and A. H. Zewail
Proc. Natl. Acad. Sci. U.S.A. **104**, 18409 (2007)

492. Structural Preablation Dynamics of Graphite Observed by Ultrafast Electron Crystallography.
F. Carbone, P. Baum, P. Rudolf, and A. H. Zewail
Phys. Rev. Lett. **100**, 035501 (2008)

493. 4D Electron Imaging: Principles and Perspectives.
D. Shorokhov and A. H. Zewail
Phys. Chem. Chem. Phys. **10**, 2879 (2008)

494. Ultrashort Electron Pulses for Diffraction, Crystallography and Microscopy: Theoretical and Experimental Resolutions.
A. Gahlmann, S. T. Park, and A. H. Zewail
Phys. Chem. Chem. Phys. **10**, 2894 (2008)

495. Unfolding and Melting of DNA (RNA) Hairpins: The Concept of Structure- Specific 2D Dynamic Landscapes.
M. M. Lin, L. Meinhold, D. Shorokhov, and A. H. Zewail
Phys. Chem. Chem. Phys. **10**, 4227 (2008)

496. Physical Biology: 4D Visualization of Complexity.
A. H. Zewail
Physical Biology: From Atoms to Medicine, ed. A. H. Zewail, Imperial College Press, London, 2008, p. 23

497. Direct Observation of the Primary Bond-Twisting Dynamics of Stilbene Anion Radical.
I-R. Lee, L. Bañares, and A. H. Zewail
J. Am. Chem. Soc. **130**, 6708 (2008)

498. 4D Visualization of Embryonic, Structural Crystallization by Single-Pulse Microscopy.
O.-H. Kwon, B. Barwick, H. S. Park, J. S. Baskin, and A. H. Zewail
Proc. Natl. Acad. Sci. U.S.A. **105**, 8519 (2008)

499. Femtosecond Diffraction with Chirped Electron Pulses.
P. Baum and A. H. Zewail
Chem. Phys. Lett. **462**, 14 (2008)

500. 4D Electron Diffraction Reveals Correlated Unidirectional Behavior in Zinc Oxide Nanowires.
D.-S. Yang, C. Lao, and A. H. Zewail
Science **321**, 1660 (2008)

501. 4D Imaging of Transient Structures and Morphologies in Ultrafast Electron Microscopy.
B. Barwick, H. S. Park, O.-H. Kwon, J. S. Baskin, and A. H. Zewail
Science **322**, 1227 (2008)

502. Nanoscale Mechanical Drumming Visualized by 4D Electron Microscopy.
O.-H. Kwon, B. Barwick, H. S. Park, J. S. Baskin, and A. H. Zewail
Nano Lett. **8**, 3557 (2008)

503. Dynamics of Ligand Substitution in Labile Cobalt Complexes Resolved by Ultrafast T-jump.
H. Ma, C. Wan, and A. H. Zewail
Proc. Natl. Acad. Sci. U.S.A. **105**, 12754 (2008)

504. Ultrafast Electron Microscopy: Watching Atoms Move and Crystals Melt.
G. K. Drayna and D. J. Flannigan; Mentor: A. H. Zewail
Caltech Undergrad Res. J. **8**, 36 (2008)

505. Reply to Catalán: Double-Proton-Transfer Dynamics of Photo-Excited 7-azaindole Dimers.
O.-H. Kwon and A. H. Zewail
Proc. Natl. Acad. Sci. U.S.A. **105**, E79 (2008)
See also:
Comment on "On the Doubly Hydrogen Bonded Dimer of 7-azaindole (0.1 M) as a Model for DNA Base Pairs in Acetonitrile Solutions at Rt" by J. Catalán. O.-H. Kwon and A. F. Mohammed
Nature Precedings, hdl:10101/npre.2008.2522.1, Posted 16 Nov 2008

506. Ultrafast Electron Diffraction Reveals Dark Structures of the Biological Chromophore Indole.
S. T. Park, A. Gahlmann, Y. He, J. S. Feenstra, and A. H. Zewail
Angew. Chem., Int. Ed. Engl. **47**, 9496 (2008)

507. Structure of Isolated Biomolecules by Electron Diffraction – Laser Desorption: Uracil and Guanine.
A. Gahlmann, S. T. Park, and A. H. Zewail
J. Am. Chem. Soc. **131**, 2806 (2009)

508. Chemistry at a Historic Crossroads.
A. H. Zewail
Chem. Phys. Chem. **10**, 23 (2009)

509. Electron and X-ray Methods of Ultrafast Structural Dynamics: Advances and Applications.
M. Chergui and A. H. Zewail
Chem. Phys. Chem. **10**, 28 (2009)

510. Direct Role of Structural Dynamics in Electron-Lattice Coupling of Superconducting Cuprates.
F. Carbone, D.-S. Yang, E. Giannini, and A. H. Zewail
Proc. Natl. Acad. Sci. U.S.A. **105**, 20161 (2008)

511. Conformations and Coherences in Structure Determination by Ultrafast Electron Diffraction.
M. M. Lin, D. Shorokhov, and A. H. Zewail
J. Phys. Chem. A **113**, 4075 (2009)

512. EELS Femtosecond Resolved in 4D Ultrafast Electron Microscopy.
F. Carbone, B. Barwick, O.-H. Kwon, H. S. Park, J. S. Baskin, and A. H.Zewail
Chem. Phys. Lett. **468**, 107 (2009)

513. Ordered Water Structure at Hydrophobic Graphite Interfaces Observed by 4D, Ultrafast Electron Crystallography.
D.-S. Yang and A. H. Zewail
Proc. Natl. Acad. Sci. U.S.A. **106**, 4122 (2009)

514. Nanomechanical Motions of Cantilevers: Direct Imaging in Real Space and Time with 4D Electron Microscopy.
D. J. Flannigan, P. C. Samartzis, A. Yurtsever, and A. H. Zewail
Nano Lett. **9**, 875 (2009)

515. Primary Peptide Folding Dynamics Observed with Ultrafast Temperature Jump.
O. F. Mohammed, G. S. Jas, M. M. Lin, and A. H. Zewail
Angew. Chem., Int. Ed. Engl. **48**, 5628 (2009)

516. Dynamics of Chemical Bonding Mapped by Energy-Resolved 4D Electron Microscopy.
F. Carbone, O.-H. Kwon, and A. H. Zewail
Science **325**, 181 (2009)

517. Temporal Lenses for Attosecond and Femtosecond Electron Pulses.
S. A. Hilbert, C. Uiterwaal, B. Barwick, H. Batelaan, and A. H. Zewail
Proc. Natl. Acad. Sci. U.S.A. **106**, 10558 (2009)

518. Charge Transfer Assisted by Collective Hydrogen-Bonding Dynamics.
O. F. Mohammed, O.- H. Kwon, C. M. Othon, and A. H. Zewail
Angew. Chem., Int. Ed. Engl. **48**, 6251 (2009)

519. Solvation in Protein (Un)folding of Melittin Tetramer–Monomer Transition.
C. M. Othon, O.-H. Kwon, M. M. Lin, and A. H. Zewail
Proc. Natl. Acad. Sci. USA **106**, 12593 (2009)

520. Structural Ultrafast Dynamics of Macromolecules: Diffraction of Free DNA and Effect of Hydration.
M. M. Lin, D. Shorokhov, and A. H. Zewail
Phys. Chem. Chem. Phys. **11**, 10619 (2009)

521. Heating and Cooling Dynamics of Carbon Nanotubes Observed by Temperature-Jump Spectroscopy and Electron Microscopy.
O. F. Mohammed, P. C. Samartzis, and A. H. Zewail
J. Am. Chem. Soc. **131**, 16010 (2009)

522. 4D Ultrafast Electron Microscopy: Imaging of Atomic Motions, Acoustic Resonances, and Moiré Fringe Dynamics.
H. S. Park, J. S. Baskin, B. Barwick, O.-H. Kwon, and A. H. Zewail
Ultramicroscopy **110**, 7 (2009)

523. Direct Observation of Martensitic Phase-Transformation Dynamics in Iron by 4D Single-Pulse Electron Microscopy.
H. S. Park, O.-H. Kwon, J. S. Baskin, B. Barwick, and A. H. Zewail
Nano Lett. **9**, 3954 (2009)

524. 4D Nanoscale Diffraction Observed by Convergent-Beam Ultrafast Electron Microscopy.

A. Yurtsever and A. H. Zewail
Science **326**, 708 (2009)

525. Photon-Induced Near-Field Electron Microscopy.
B. Barwick, D. J. Flannigan, and A. H. Zewail
Nature **462**, 902 (2009)

526. New Light on Molecular and Materials Complexity: 4D
Electron Imaging.
D. Shorokhov and A. H. Zewail
J. Am. Chem. Soc. **131**, 17998 (2009)

527. 4D Electron Microscopy.
A. H. Zewail
Science **328**, 187 (2010)

528. Micrographia of the 21st Century: From Camera
Obscura to 4D Microscopy.
A. H. Zewail
Phil. Trans. R. Soc. A **368**, 1191 (2010)

529. The New Age of Structural Dynamics.
A. H. Zewail
Acta Cryst. A **66**, 135 (2010)

530. Filming the Invisible in 4D.
A. H. Zewail
Sci. Am. **303**, 74 (2010)

531. 4D Attosecond Imaging with Free Electrons: Diffraction
Methods and Potential Applications.
P. Baum and A. H. Zewail
Chem. Phys. **366**, 2 (2009)

532. Real-Time Observation of Cuprates Structural
Dynamics by Ultrafast Electron Crystallography.
F. Carbone, N. Gedik, J. Lorenzana, and A. H. Zewail
Adv. Cond. Mat. Phys. **2010**, 958618 (2010)

533. Four-dimensional Visualization of Transitional Structures in Phase Transformations by Electron Diffraction.
P. Baum, D.-S. Yang, and A. H. Zewail
Springer Series in Chemical Physics, Vol. 92: Ultrafast Phenomena XVI, eds. P. Corkum, S. De Silvestri, K. A. Nelson, E. Riedle, and R. W. Schoenlein, Springer-Verlag, Berlin-Heidelberg, 2009, p. 116

534. Attosecond Free Electron Pulses for Diffraction and Microscopy.
P. Baum and A. H. Zewail
Springer Series in Chemical Physics, Vol. 92: Ultrafast Phenomena XVI, eds. P. Corkum, S. De Silvestri, K. A. Nelson, E. Riedle, and R. W. Schoenlein, Springer-Verlag, Berlin-Heidelberg, 2009, p. 155

535. Optomechanical and Crystallization Phenomena Visualized with 4D Electron Microscopy: Interfacial Carbon Nanotubes on Silicon Nitride.
D. J. Flannigan and A. H. Zewail
Nano Lett. **10**, 1892 (2010)

536. Structural Dynamics and Transient Electric-Field Effects in Ultrafast Electron Diffraction from Surfaces.
S. Schäfer, W. Liang, and A. H. Zewail
Chem. Phys. Lett. **493**, 11 (2010)

537. Biological Imaging with 4D Ultrafast Electron Microscopy.
D. J. Flannigan, B. Barwick, and A. H. Zewail
Proc. Natl. Acad. Sci. U.S.A. **107**, 9933 (2010)

538. 4D Electron Tomography.
O.-H. Kwon and A. H. Zewail
Science **328**, 1668 (2010)

539. Ultrafast Electronic and Structural Phenomena in
     Graphite and Graphene.
     F. Carbone, O.-H. Kwon, A. Cannizzo, F. van Mourik,
     M. Chergui, and A. H. Zewail
     Proc. 1st Int. Conf. on Ultrafast Structural Dynamics,
     eds. M. Chergui, A. Cannizzo, and G. Auböck, EPFL,
     Lausanne, 2010, p. 28

540. Femtochemistry, Femtobiology and Femtophysics.
     A. H. Zewail
     Sci. China G **53**, 976 (2010)

541. Nonchaotic, Nonlinear Motion Visualized in Complex
     Nanostructures by Stereographic 4D Electron
     Microscopy.
     O.-H. Kwon, H. S. Park, J. S. Baskin, and A. H. Zewail
     Nano Lett. **10**, 3190 (2010)

542. Direct Structural Determination of Conformations of
     Photoswitchable Molecules by Laser Desorption –
     Electron Diffraction.
     A. Gahlmann, I-R. Lee, and A. H. Zewail
     Angew. Chem. Int. Ed. Engl. **49**, 6524 (2010)

543. Scanning Ultrafast Electron Microscopy.
     D.-S. Yang, O. F. Mohammed, and A. H. Zewail
     Proc. Natl. Acad. Sci. U.S.A. **107**, 14993 (2010)

544. Hydration Dynamics at Fluorinated Protein Surfaces.
     O.-H. Kwon, T. H. Yoo, C. M. Othon, J. A. Van
     Deventer, D. A. Tirrell, and A. H. Zewail
     Proc. Natl. Acad. Sci. U.S.A. **107**, 17101 (2010)

545. 4D Lorentz Electron Microscopy Imaging: Magnetic
     Domain Wall Nucleation, Reversal, and Wave
     Velocity.
     H. S. Park, J. S. Baskin, and A. H. Zewail
     Nano Lett. **10**, 3796 (2010)

546. Nanofriction Visualized in Space and Time by 4D Electron Microscopy.
D. J. Flannigan, S. T. Park, and A. H. Zewail
Nano Lett. **10**, 4767 (2010)

547. Photon-Induced Near-Field Electron Microscopy (PINEM): Theoretical and Experimental.
S. T. Park , M. M. Lin, and A. H. Zewail
New J. Phys. **12**, 123028 (2010)

548. The Future of Chemical Physics.
A. H. Zewail
Chem. Phys. **378**, 1 (2010)

549. Biological Water: A Critique.
D. Zhong, S. K. Pal, and A. H. Zewail
Chem. Phys. Lett. **503**, 1 (2011)

550. Kikuchi Ultrafast Nanodiffraction in Four-Dimensional Electron Microscopy.
A. Yurtsever and A. H. Zewail
Proc. Natl. Acad. Sci. U.S.A. **108**, 3152 (2011)

551. Irreversible Chemical Reactions Visualized in Space and Time with 4D Electron Microscopy.
S. T. Park, D. J. Flannigan, and A. H. Zewail
J. Am. Chem. Soc. **133**, 1730 (2011)

552. Macromolecular Structural Dynamics Visualized by Pulsed Dose Control in 4D Electron Microscopy.
O.-H. Kwon, V. Ortalan, and A. H. Zewail
Proc. Natl. Acad. Sci. U.S.A. **108**, 6026 (2011)

553. Nanomusical Systems Visualized and Controlled in 4D Electron Microscopy.
J. S. Baskin, H. S. Park, and A. H. Zewail
Nano Lett. **11**, 2183 (2011)

554. Primary Structural Dynamics in Graphite.
S. Schäfer, W. Liang, and A. H. Zewail
New. J. Phys. **13**, 063030 (2011)

555. 4D Scanning Ultrafast Electron Microscopy:
Visualization of Materials Surface Dynamics.
O. F. Mohammed, D.-S. Yang, S. K. Pal, and A. H. Zewail
J. Am. Chem. Soc. **133**, 7708 (2011)

556. 4D Scanning Transmission Ultrafast Electron
Microscopy: Single-Particle Imaging and Spectroscopy.
V. Ortalan and A. H. Zewail
J. Am. Chem. Soc. **133**, 10732 (2011)

557. Structural Dynamics of Free Proteins in Diffraction.
M. M. Lin, D. Shorokhov, and A. H. Zewail
J. Am. Chem. Soc. **133**, 17072 (2011)

558. Structural Dynamics of Free Amino Acids in
Diffraction.
I-R. Lee, A. Gahlmann, and A. H. Zewail
Angew. Chem. Int. Ed. Engl. **51**, 99 (2012)

559. Speed Limit of Protein Folding Evidenced in Secondary
Structure Dynamics.
M. M. Lin, O. F. Mohammed, G. S. Jas, and A. H. Zewail
Proc. Natl. Acad. Sci. U.S.A. **108**, 16622 (2011)

560. Structural Dynamics of Nanoscale Gold by Ultrafast
Electron Crystallography.
S. Schäfer, W. Liang, and A. H. Zewail
Chem. Phys. Lett. **515**, 278 (2011)

561. Structural Dynamics of Surfaces by Ultrafast Electron
Crystallography: Experimantal and Multiple Scattering
Theory.
S. Schäfer, W. Liang, and A. H. Zewail
J. Chem. Phys. **135**, 214201 (2011)

562. Enhancing Image Contrast and Slicing Electron Pulses in 4D Near-Field Electron Microscopy.
S. T. Park and A. H. Zewail
Chem. Phys. Lett. **521**, 1 (2012)

563. Conical Intersections: Theory, Computation and Experiment.
A. H. Zewail
*Advanced Series in Physical Chemistry, Vol. 17;
Conical Intersections: Theory, Computation and
Experiment*, eds. W. Domcke, D. R. Yarkony,
and H. Köppel, World Scientific, Singapore, 2011, p. v

564. Nobel Prize Winners in Chemistry.
A. H. Zewail
*Nobel Prize Winners in Chemistry, 1901–2009:
Nobelists in Chemistry*, eds. D. N. Dhar and P. Dhar,
Lambert Academic, Saarbrücken, 2011, p. iv

565. Subparticle Ultrafast Spectrum Imaging in 4D Electron Microscopy.
A. Yurtsever, R. M. van der Veen, and A. H. Zewail
Science **335**, 59 (2012)

566. Chirped Imaging Pulses in Four-Dimensional Electron Microscopy: Femtosecond Pulsed Hole Burning.
S. T. Park, O.-H. Kwon, and A. H. Zewail
New J. Phys. **14**, 053046 (2012)

567. 4D Electron Microscopy Visualization of Anisotropic Atomic Motions in Carbon Nanotubes.
S. T. Park, D. J. Flannigan, and A. H. Zewail
J. Am. Chem. Soc. **134**, 9146 (2012)

568. Ultrafast Electron Crystallography of Monolayer Adsorbates on Clean Surfaces: Structural Dynamics.
W. Liang, S. Schäfer, and A. H. Zewail
Chem. Phys. Lett. **542**, 1 (2012)

569. Ultrafast Electron Crystallography of Heterogeneous
Structures: Gold-Graphene Bilayer and Ligand-
Encapsulated Nanogold on Graphene.
W. Liang, S. Schäfer, and A. H. Zewail
Chem. Phys. Lett. **542**, 8 (2012)

570. Direct Visualization of Near-Fields in Nanoplasmonics
and Nanophotonics.
A. Yurtsever and A. H. Zewail
Nano Lett. **12**, 3334 (2012)

571. Ultrafast Kikuchi Diffraction: Nanoscale Stress-Strain
Dynamics of Wave-Guiding Structures.
A. Yurtsever, S. Schäfer, and A. H. Zewail
Nano Lett. **12**, 3772 (2012)

572. Hydrophobic Forces and the Length Limit of Foldable
Protein Domains.
M. M. Lin and A. H. Zewail
Proc. Natl. Acad. Sci. U.S.A. **109**, 9851 (2012)

573. Protein Folding—Simplicity in Complexity.
M. Lin and A. H. Zewail
Ann. Phys. **524**, 379 (2012)

574. Entangled Nanoparticles: Discovery by Visualization in
4D Electron Microscopy.
A. Yurtsever, J. S. Baskin and A. H. Zewail
Nano Lett. **12**, 5027 (2012)

575. Relativistic Effects in Photon-Induced Near Field
Electron Microscopy.
S. T. Park and A. H. Zewail
J. Phys. Chem. A **116**, 11128 (2012)

576. 4D Electron Microscopy: Principles and Applications.
D. Flannigan and A. H. Zewail
Acc. Chem. Res. **45**, 1828 (2012)

577. Environmental Scanning Ultrafast Electron Microscopy: Structural Dynamics of Solvation at Interfaces.
D.-S. Yang, O. F. Mohammed, and A. H. Zewail
Angew. Chem. Int. Ed. Engl. **52**, 2897 (2013)

578. Biomechanics of DNA Structures Visualized by 4D Electron Microscopy.
U. J. Lorenz and A. H. Zewail
Proc. Natl. Acad. Sci. U.S.A. **110**, 2822 (2013)

579. Single-Nanoparticle Phase Transitions Visualized by 4D Electron Microscopy.
R. M. van der Veen, O.-H. Kwon, A. Tissot, A. Hauser, and A. H. Zewail
Nature Chem. **5**, 395 (2013)

580. Unusual Molecular Material Formed through Irreversible Transformation and Revealed by 4D Electron Microscopy.
R. M. van der Veen, A. Tissot, A. Hauser, and A. H. Zewail
Phys. Chem. Chem. Phys. **15**, 7831 (2013)

581. Exceptional Rigidity and Biomechanics of Amyloid Revealed by 4D Electron Microscopy.
A. W. P. Fitzpatrick, S. T. Park, and A. H. Zewail
Proc. Natl. Acad. Sci. U.S.A. **110**, 10976 (2013)

582. 4D Cryo-Electron Microscopy of Proteins.
A. W. P. Fitzpatrick, U. J. Lorenz, G. M. Vanacore, and A. H. Zewail
J. Am. Chem. Soc. **135**, 19123 (2013).

583. Graphene-Layered Steps and Their Fields Visualized by 4D Electron Microscopy.
S. T. Park, A. Yurtsever, J. S. Baskin, and A. H. Zewail
Proc. Natl. Acad. Sci. U.S.A. **110**, 9277 (2013)

584. Visualization of Carrier Dynamics in p(n)-Type GaAs by Scanning Ultrafast Electron Microscopy.
J. Cho, T. Y. Hwang, and A. H. Zewail
Proc. Natl. Acad. Sci. U.S.A. **111**, 2094 (2014)

585. 4D Imaging and Diffraction Dynamics of Single-Particle Phase Transition in Heterogeneous Ensembles.
H. Liu, O.-H. Kwon, J. Tang, and A. H. Zewail
Nano Lett. **14**, 946 (2014)

586. Structural Dynamics Effects on the Ultrafast Chemical Bond Cleavage of a Photodissociation Reaction.
M. E. Corrales, V. Loriot, G. Balerdi, J. González-Vázquez, R. de Nalda, L. Bañares, and A. H. Zewail
Phys. Chem. Chem. Phys. **16**, 8812 (2014)

587. Seeing in 4D with Electrons: Development of Ultrafast Electron Microscopy at Caltech.
J. S. Baskin and A. H. Zewail
Compt. Rend. Phys. **15**, 176 (2014)

588. Photon-Induced Near-Field Electron Microscopy: Mathematical Formulation of the Relation between the Experimental Observables and the Optically Driven Charge Density of Nanoparticles.
S. T. Park and A. H. Zewail
Phys. Rev. A **89**, 013851 (2014)

589. Observing (Non)linear Lattice Dynamics in Graphite by Ultrafast Kikuchi Diffraction.
W. Liang, G. M. Vanacore, and A. H. Zewail
Proc. Natl. Acad. Sci. U.S.A. **111**, 5491 (2014)

590. Observing Liquid Flow in Nanotubes by 4D Electron Microscopy.
U. J. Lorenz and A. H. Zewail
Science **344**, 1496 (2014)

591. 4D Multiple-Cathode Ultrafast Electron Microscopy.
J. S. Baskin, H. Liu, and A. H. Zewail
Proc. Natl. Acad. Sci. U.S.A. **111**, 10479 (2014)

592. Dominance of Misfolded Intermediates in the Dynamics
of α-Helix Folding.
M. M. Lin, D. Shorokhov, and A. H. Zewail
Proc. Natl. Acad. Sci. U.S.A. **111**, 14424 (2014)

593. Characterization of Fast Photoelectron Packets in Weak
and Strong Laser Fields in Ultrafast Electron
Microscopy.
D. A. Plemmons, S. T. Park, A. H. Zewail, and
D. J. Flannigan
Ultramicroscopy **146**, 97 (2014)

594. Diffraction of Quantum Dots Reveals Nanoscale
Ultrafast Energy Localization.
G. M. Vanacore, J. Hu, W. Liang, S. Bietti, S.
Sanguinetti, and A. H. Zewail
Nano Lett. **14**, 6148 (2014)

595. Four-Dimensional Imaging of Carrier Interface
Dynamics in p-n Junctions.
E. Najafi, T. D. Scarborough, J. Tang, and A. H. Zewail
Science **347**, 164 (2015)

596. Origin of Axial and Radial Expansions in Carbon
Nanotubes Revealed by Ultrafast Diffraction and
Spectroscopy.
G. M. Vanacore, R. M. van der Veen, and A. H. Zewail
ACS Nano **9**, 1721 (2015)

597. Ultrafast Core-Loss Spectroscopy in Four-Dimensional
Electron Microscopy.
R. M. van der Veen, T. J. Penfold, and A. H. Zewail
Struct. Dyn. **2**, 024302 (2015)

598. Nanomechanics and Intermolecular Forces of Amyloid Revealed by Four- Dimensional Electron Microscopy.
A. W. P. Fitzpatrick, G. M. Vanacore, and A. H. Zewail
Proc. Natl. Acad. Sci. U.S.A. **112**, 3380 (2015)

599. Photon Gating in 4D Electron Microscopy.
M. T. Hassan, H. Liu, J. S. Baskin, and A. H. Zewail
Proc. Natl. Acad. Sci. U.S.A. **112**, 12944 (2015)

600. Photonics and Plasmonics in 4D Ultrafast Electron Microscopy.
B. Barwick and A. H. Zewail
ACS Photon. **2**, 1391 (2015)

601. Observing in Space and Time the Ephemeral Nucleation of Liquid-to-Crystal Phase.
B.-K. Yoo, O.-H. Kwon, H. Liu, J. Tang, and A. H. Zewail
Nature Commun. **6**, 8639 (2015)

602. Transient Structures and Possible Limits of Data Recording in Phase-Change Materials.
J. Hu, G. M. Vanacore, Z. Yang, X. Miao, and A. H. Zewail
ACS Nano **9**, 6728 (2015)

603. On the Dynamical Nature of the Active Center in a Single-Site Photocatalyst Visualized by 4D Ultrafast Electron Microscopy.
B.-K. Yoo, Z. Su, J. M. Thomas, and A. H. Zewail
Proc. Natl. Acad. Sci. U.S.A. **113**, 503 (2016)

604. Infrared PINEM Developed by Diffraction in 4D UEM.
H. Liu, J. S. Baskin, and A. H. Zewail
Proc. Natl. Acad. Sci. U.S.A. **113**, 2041 (2016)

605. 4D Ultrafast Electron Microscopy: Evolutions and Revolutions.
D. Shorokhov and A. H. Zewail
J. Chem. Phys. **144**, 080901 (2016)

606. Four-dimensional Electron Microscopy: Ultrafast Imaging, Diffraction and Spectroscopy in Materials Science and Biology. G. M. Vanacore, A. W. P. Fitzpatrick, and A. H. Zewail *Nano Today* 11, 228 (2016)

11.2 Dr. Zewail's Patents and Books

From the list of patents he filed, one can determine when he got his new scientific ideas, discover his most productive period and identify the people who shared his success, from students to colleagues.

11.2.1 *A list of Dr. Zewail's patents*

1. *Luminescent Solar Energy Concentrator Devices*, A. H. Zewail and J. S. Batchelder, California Institute of Technology, *U.S. Pat.* 4,227,939, October 14, 1980

2. *Method and System for Ultrafast Photoelectron Microscope*, A. H. Zewail and V. A. Lobastov, California Institute of Technology, *U.S. Pat.* 7,154,091, December 26, 2006

3. *4D Imaging in an Ultrafast Electron Microscope*, A. H. Zewail, California Institute of Technology, *U.S. Pat.* 8,203,120, June 19, 2012

4. *Characterization of Nanoscale Structures Using an Ultrafast Electron Miscroscope*, A. H. Zewail, California Institute of Technology, *U.S. Pat.* 8,247,769, August 21, 2012; Continued, *U.S. Pat.* 8,440,970, May 14, 2013; Continued, *U.S. Pat.* 8,686,359, April 1, 2014; Continued, *U.S. Pat.* 9,053,903, June 9, 2015

5. *Photon Induced Near Field Electron Microscope and Biological Imaging System*, A. H. Zewail, D. J. Flannigan and B. Barwick, California Institute of Technology, *U.S. Pat.* 8,429,761, April 23, 2013; Continued, *U.S. Pat.* 8,569,695, October 29, 2013; Continued, *U.S. Pat.* 8,963,085, February 24, 2015

6. *Control Imaging Methods in Advanced Ultrafast Electron Microscopy*, A. H. Zewail, J. S. Baskin, California Institute of Technology, *U.S. Pat.* 8,766,181, July 1, 2014

7. *Method and System for 4D Tomography and Ultrafast Scanning Electron Microscopy*, A. H. Zewail, O.-H. Kwon, O. F. Mohammed Abdelsaboor, D. S. Yang, California Institute of Technology, *U.S. Pat.* 8,841,613, September 23, 2014

11.2.2 *Non-scientific books published by Dr. Zewail*

1. *Voyage Through Time: Walks of Life to the Nobel Prize*, A. H. Zewail, American University in Cairo (AUC), Cairo, 2002
2. *Age of Science (Asr Al Álm, in Arabic)*, A. H. Zewail, Dar Al Shorouk, Beirut-Cairo, 2005
3. *Time (Al Zaman, in Arabic)*, Book Series, A. H. Zewail, Dar Al Shorouk, Cairo, 2007
4. *Dialogue of Civilizations (Hewar Al Hadarat, in Arabic)*, Book Series, A. H. Zewail, Dar Al Shorouk, Cairo, 2007
5. *Reflections on World Affairs: Peace and Politics*, Essential Reading Series on Eastern Culture, Civilization and History, Vol. 1, A. H. Zewail, Imperial College Press, London, 2015

Dr. Zewail was the initiator of a social society for intellectuals in Egypt (Asr El Elm) which was registered in the year 2005. It started with Abdelmeguid Khalid, the Chemistry teacher at Mansoura Girls College, as vice-chairman to Dr. Zewail.

The list of books published by Dr. Zewail clearly shows his growing interest in politics. There are some passages in his books which show how proud he was to be Egyptian. Such emotions motivated him to serve his nation. Of course he gained a lot of awards, but he did not pursue them. With Dr. Zewail, his country always came first.

11.2.3 *Scientific books published by Dr. Zewail*

1. *Advances in Laser Spectroscopy I*, A. H. Zewail, SPIE, Bellingham, 1977
2. *Advances in Laser Chemistry*, ed. A. H. Zewail, Springer-Verlag, Berlin-Heidelberg, 1978

3. *Photochemistry and Photobiology*, Vols. 1 and 2, ed. A. H. Zewail, Harwood Academic, London, 1983

4. *Ultrafast Phenomena VII*, C. B. Harris, E. P. Ippen, G. A. Mourou and A. H. Zewail, Springer-Verlag, Berlin-Heidelberg, 1990

5. *The Chemical Bond: Structure and Dynamics*, A. H. Zewail, Academic Press, Boston, 1992

6. *Ultrafast Phenomena VIII*, J.-L. Martin, A. Migus, G. A. Mourou and A. H. Zewail, Springer-Verlag, Berlin-Heidelberg, 1993

7. *Ultrafast Phenomena IX, eds.* P. F. Barbara, W. H. Knox, G. A. Mourou and A. H. Zewail, Springer-Verlag, Berlin-Heidelberg, 1994

8. *Femtochemistry: Ultrafast Dynamics of the Chemical Bond, Vols. 1 and 2*, A. H. Zewail, World Scientific, Singapore, 1994

9. *Physical Biology: From Atoms to Medicine*, A. H. Zewail, Imperial College Press, London, 2008

10. *4D Electron Microscopy: Imaging in Space and Time*, A. H. Zewail and J. M. Thomas, Imperial College Press, London, 2010

11. *4D Visualization of Matter: Recent Collected Works*, A. H. Zewail, Imperial College Press, London, 2014

11.3 Dr. Zewail's awards and honors

Special Honors

- King Faisal International Prize in Science (1989).
- First Linus Pauling Chair, Caltech (1990).
- Wolf Prize in Chemistry (1993).
- Order of Merit, first class (Sciences & Arts), from President Mubarak (1995).
- Robert A. Welch Award in Chemistry (1997).
- Benjamin Franklin Medal, Franklin Institute, USA (1998).
- Egypt Postage Stamps, with Portrait (1998); the Fourth Pyramid (1999).
- Nobel Prize in Chemistry (1999).

- Grand Collar of the Nile, Highest State Honor, conferred by President Mubarak (1999).

Prizes and Awards

- Alfred P. Sloan Foundation Fellow (1978–1982).
- Camille and Henry Dreyfus Teacher-Scholar Award (1979–1985).
- Alexander von Humboldt Award for Senior United States Scientists (1983).
- National Science Foundation Award for especially creative research (1984; 1988;1993).
- Buck-Whitney Medal, American Chemical Society (1985).
- John Simon Guggenheim Memorial Foundation Fellow (1987).
- Harrison Howe Award, American Chemical Society (1989).
- Topical Society of Laser Sciences, TSLS Cairo University Golden Medal (1991)
- Cairo University Shield of Excellence (1991).
- Carl Zeiss International Award, Germany (1992).
- Earle K. Plyler Prize, American Physical Society (1993).
- Medal of the Royal Netherlands Academy of Arts and Sciences, Holland (1993).
- Bonner Chemiepreis, Germany (1994).
- Cairo University Grand Shield, NILES (1994).
- Herbert P. Broida Prize, American Physical Society (1995).
- Leonardo Da Vinci Award of Excellence, France (1995).
- Collége de France Medal, France (1995).
- Peter Debye Award, American Chemical Society (1996).
- National Academy of Sciences Award, Chemical Sciences, USA (1996).
- J.G. Kirkwood Medal, Yale University (1996).
- Peking University Medal, PU President, Beijing, China (1996).
- Pittsburgh Spectroscopy Award (1997).
- First E.B. Wilson Award, American Chemical Society (1997).
- Linus Pauling Medal Award (1997).

- Richard C. Tolman Medal Award (1998).
- William H. Nichols Medal Award (1998).
- Paul Karrer Gold Medal, University of Zürich, Switzerland (1998).
- E.O. Lawrence Award, U.S. Government (1998).
- Merski Award, University of Nebraska (1999).
- Röntgen Prize, (100th Anniversary of the Discovery of X-rays), Germany (1999).

Academies and Societies

- American Physical Society, Fellow (elected 1982).
- National Academy of Sciences, USA (elected 1989).
- Third World Academy of Sciences, Italy (elected 1989).
- Sigma Xi Society, USA (elected 1992).
- American Academy of Arts and Sciences (elected 1993).
- Académie Européenne des Sciences, des Arts et des Lettres, France (elected 1994).
- American Philosophical Society (elected 1998).
- Pontifical Academy of Sciences (elected 1999).
- American Academy of Achievement (elected 1999).
- Royal Danish Academy of Sciences and Letters (elected 2000)
- MTPR-004. Faculty of Science, Cairo University (2004)

Honorary Degrees

- Oxford University, UK (1991): M.A., h.c.
- American University, Cairo, Egypt (1993): D.Sc., h.c.
- Katholieke Universiteit, Leuven, Belgium (1997): D.Sc., h.c.
- University of Pennsylvania, USA (1997): D.Sc., h.c.
- Université de Lausanne, Switzerland (1997): D.Sc., h.c.
- Swinburne University, Australia (1999): D.U., h.c.
- Arab Academy for Science & Technology, Egypt (1999): H.D.A.Sc.
- Alexandria University, Egypt (1999): H.D.Sc.
- University of New Brunswick, Canada (2000): Doctoris in Scientia, D.Sc., h.c.

- Universita di Roma "La Sapienza", Italy (2000): D.Sc., h.c.
- Université de Liège, Belgium (2000): Doctor honoris causa, D., h.c.
- From Les Prix Nobel. The Nobel Prizes 1999, Editor Tore Frängsmyr, Nobel Foundation, Stockholm, 2000.

We pray for Allah, the merciful and the great, to reward Ahmed Zewail with paradise after his life on earth.